Structure-Properties-Processing Relationships in Metallic Materials

Structure-Properties-Processing Relationships in Metallic Materials

Editor

Spyros Papaefthymiou

MDPI • Basel • Beijing • Wuhan • Barcelona • Belgrade • Manchester • Tokyo • Cluj • Tianjin

Editor
Spyros Papaefthymiou
School of Mining &
Metallurgical Engineering
Greece

Editorial Office
MDPI
St. Alban-Anlage 66
4052 Basel, Switzerland

This is a reprint of articles from the Special Issue published online in the open access journal *Metals* (ISSN 2075-4701) (available at: https://www.mdpi.com/journal/metals/special_issues/structure_properties_processing_metals).

For citation purposes, cite each article independently as indicated on the article page online and as indicated below:

LastName, A.A.; LastName, B.B.; LastName, C.C. Article Title. *Journal Name* **Year**, *Volume Number*, Page Range.

ISBN 978-3-0365-3597-5 (Hbk)
ISBN 978-3-0365-3598-2 (PDF)

Cover image courtesy of Spyros Papaefthymiou (GE) & Sofia Papadopoulou

Contents

About the Editor

Spyros Papaefthymiou is currently an Associate Professor (since 2020) in the Laboratory of Physical Metallurgy in the School of Mining and Metallurgical Engineering of the National Technical University of Athens (NTUA) in Greece. His interests include metal forming/processing, welding–joining, physical and numerical modelling, and simulation. In addition to these topics, he deals with sustainable development and efficient and modern teaching methods. He is a mining and metallurgical engineer, who graduated from the NTUA (in Greek: "Ethniko Metsovio Polytechnio") in 2000 and received the academic title of "Dr.-Ing." in 2005 from RWTH Aachen University, Germany. As a researcher in Aachen and Delft (at the Technical University of Delft, The Netherlands) he focused on advanced high-strength steels (AHSS) for the automotive industry, e.g., TRIP transformation-induced plasticity (TRIP), dual-phase (DP), and complex-phase (CP) steels. He works in the field of physical metallurgy and alloy design and, more specifically, specializes in microstructure–properties–fracture relationships, the forming of metals and alloys, welding, joining and heat treatments, the mechanical properties of materials and the modelling and simulation of microstructures and materials processing. He has published 133 articles (48 papers in international journals, 45 articles in international conference proceedings and 40 in national conference proceedings). He wrote the book *Materials Technology* (in Greek) (ISBN: 978-960-337-129-8, Eugenides Foundation, 2nd Edition, 2017) and a chapter about industrial pipeline welding in a collective book published by Springer (Papaefthymiou S. Industrial Pipeline Welding. In *Welding Technology. Materials Forming, Machining and Tribology*; Davim, J.P., ed.; Springer International Publishing: Cham, Switzerland, 2021; pp. 387–413 https://doi.org/10.1007/9783030639860_12). Since 2014, he has completed the supervision of 64 diploma theses, 2 master's theses and 3 PhD theses; he has also co-supervised 102 diploma theses to date (he currently supervises 14 diploma theses and 4 PhD theses, and co-supervises 9 PhDs). Over the last 8 years in academia, he has introduced new pedagogic methods into teaching. He motivates students using a coaching approach. Collective learning through projects and workshops enable familiarization with the available digital tools. Students learn to present effectively, e.g., according to the "elevator moment", they understand how assessment and evaluation is a vital part of the education process. Working in groups, tackling critical tasks, following tight schedules and adapting to challenges infuses the students with valuable leadership skills. In June 2021, he joined the "Reviewers Board" of the international journal *Materials*, where high-performing reviewers are to acknowledge their assistance in the highly demanding peer-review process.

Preface to "Structure-Properties-Processing Relationships in Metallic Materials"

The steady innovation in the materials sector over the last three decades has been supported by the introduction of simulation and analytical technique tools in the development chain. Materials must be stronger, lighter, recyclable and more attractive in all aspects of their utilization. Microstructure adjustments to operational requirements were successfully introduced through optimized chemical composition and thermomechanical processing step control, based not only on experimental development, but also on physical and numerical simulation approaches. This Special Issue encompasses some of these areas, providing insights into the complexity of the research and development regarding industrial materials.

Spyros Papaefthymiou
Editor

Editorial

Structure–Properties–Processing Relationships in Metallic Materials

Spyros Papaefthymiou

School of Mining & Metallurgical Engineering, National Technical University of Athens, 15780 Athens, Greece; spapaef@metal.ntua.gr; Tel.: +30-210-772-4710

Citation: Papaefthymiou, S. Structure–Properties–Processing Relationships in Metallic Materials. *Metals* **2022**, *12*, 361. https://doi.org/10.3390/met12020361

Received: 19 January 2022
Accepted: 16 February 2022
Published: 21 February 2022

Publisher's Note: MDPI stays neutral with regard to jurisdictional claims in published maps and institutional affiliations.

1. Introduction and Scope

The steady innovation of materials has been assisted by the introduction of simulation and analytical technique tools in the development chain. The necessity for increased efficiency and power and low emissions and, in parallel, safety, has brought forth new, advanced, high and ultra-high strength multiphase steel grades into use in the automotive industry, one of the most competitive sectors in the world. The steadily stricter rules for vehicle emissions forces the broader usage of light-weight metals and alloys.

Microstructure–properties relationships in combination with processing or alloying strategies for the development of tailored microstructures and, thus, mechanical properties, in new alloys have been intensively investigated by academia and industry throughout the years. Simulation approaches for addressing not only diffusional but also shear and displacive transformations are now of great interest for process simulation and the control of microstructure evolution, taking into consideration processing conditions and/or limitations. Alloys with a need for texture-oriented rolling and/or with tailored rolling and heat treatment schedules underline the necessity for proper structure–properties–processing control.

2. Contributions

In the current Special Issue of *Metals*, we host six contributions from academia and industry on their latest research developments and achievements in their applied research field. These elucidate on the effect of temperature during forming operations, e.g., in aluminum alloys closely analyzed via texture evaluation. Additionally, the opportunities that derive from microstructure and process simulation ensure that from the deeper understanding of the microstructure evolution, a further optimization of the alloys can be achieved, taking the capabilities of the forming equipment and the reduction schedules into consideration. Furthermore, trace elements and their effect in critical properties such as steel hot ductility performance or the role of non-metallic inclusions in the mechanical behavior of cold drawn pearlitic steel are also reviewed in this Special Issue.

Papadopoulou et al. [1] show how during forming, thickness reduction and thermal treatment affect the recrystallization and evolution of the crystallographic texture of Al5182, which in turn influences the mechanical properties. Thus, by controlling the texture via proper reduction sequences, targeted r-values, yield and tensile strength can be obtained to suit further forming of the material. Gavalas and Papaefthymiou [2] clearly show how flatness, an important quality characteristic of rolled products, is tricky to be controlled by a rolling mill operator and how a proper thermo-mechanical finite element model can be of significant assistance by identifying critical process parameters (e.g., strip temperature, cooling characteristics, roll core temperature and the evolution of the thermal camber). Baganis et al. [3] apply phase field simulation for the optimization of Al6XXX heat treatment in order to help understand how microstructure evolution during heat treatment, more specifically recrystallisation kinetics, grain growth and interface mobility, affects the mechanical properties of this Al-Mg-Si alloy family.

In their work, Guo et al. [4] show the important role of trace impurities such as phosphorus and rare earth cerium on the hot ductility in the performance of advanced SA508-4N RPV steel. The addition of P and Ce is able to facilitate the occurrence of the dynamic recrystallization (DR) of the steel, lowering the initial temperature of DR from ~900 to ~850 °C and thereby enhancing the hot ductility performance. Consequently, the combined addition of P and Ce can significantly improve the hot ductility of SA508-4N RPV steel, thereby improving its continuous casting performance and hot workability [4].

Toribio et al. [5] showed the importance of detailed metallographic and scanning electron microscope (SEM) microstructure analysis of microstructural defects exhibited by pearlitic steels in order to unveil how their evolution by cold drawing has consequences on the isotropic/anisotropic fracture behavior of different steels. Such defects caused by non-metallic inclusions have a very relevant role in the fracture behavior of cold drawn pearlitic steels [5]. Papadopoulou et al. [6] also present the importance of advanced microstructure analysis in understanding microstructure evolution and optimizing mechanical behavior. The effect of thermal processing on the initial microstructure, texture evolution and anisotropy, as shown by the example of Al3104 sheets, has a paramount effect on mechanical properties [6]. The texture characteristics, i.e., the preservation of an increased amount of S components, and the presence of strain-free elongated grains along with the coexistence of a complex and resistant-to-crack-propagation substructure consisting of both high-angle grain boundaries (HAGBs) and subgrain boundaries (SGBs) can lead to an optimal combination of Δr and r_m parameters [6].

These contributions underline the importance of synergetic experimental and simulation analyses for the optimization of modern materials using thermal processing steps, rolling mills, casting machines, etc., as tools for the development of proper microstructures suitable for advanced applications.

3. Conclusions

Today, for material optimization purposes amongst traditional physical metallurgical approaches and new characterization techniques, simulation studies on microstructure–properties relationships provide great benefits for the scientific and technical evaluation of existing alloys. The optimization and the development of new alloys can be performed based on the processing capabilities (thermal and forming operational needs).

Our constant aim is to bridge academia and industry, providing the required theoretical aspects for advanced applications in the industry under the prism of microstructure–properties–processing relationships incorporating simulation (both numerical and physical) that can serve the purpose of the optimization of new materials to be applied and/or concepts and novel aspects of current or new, to-be-adopted steel grades.

Funding: This research received no external funding.

Conflicts of Interest: The author declares no conflict of interest.

References

1. Papadopoulou, S.; Kontopoulou, A.; Gavalas, E.; Papaefthymiou, S. The Effects of Reduction and Thermal Treatment on the Recrystallization and Crystallographic Texture Evolution of 5182 Aluminum Alloy. *Metals* **2020**, *10*, 1380. [CrossRef]
2. Gavalas, E.; Papaefthymiou, S. Thermal Camber and Temperature Evolution on Work Roll during Aluminum Hot Rolling. *Metals* **2020**, *10*, 1434. [CrossRef]
3. Guo, Y.; Zhao, Y.; Song, S. Highly Enhanced Hot Ductility Performance of Advanced SA508-4N RPV Steel by Trace Impurity Phosphorus and Rare Earth Cerium. *Metals* **2020**, *10*, 1598. [CrossRef]
4. Baganis, A.; Bouzouni, M.; Papaefthymiou, S. Phase Field Simulation of AA6XXX Aluminium Alloys Heat Treatment. *Metals* **2021**, *11*, 241. [CrossRef]
5. Toribio, J.; Ayaso, F.; González, B. Role of Non-Metallic Inclusions in the Fracture Behavior of Cold Drawn Pearlitic Steel. *Metals* **2021**, *11*, 962. [CrossRef]
6. Papadopoulou, S.; Loukadakis, V.; Zacharopoulos, Z.; Papaefthymiou, S. Influence of Uniaxial Deformation on Texture Evolution and Anisotropy of 3104 Al Sheet with Different Initial Microstructure. *Metals* **2021**, *11*, 1729. [CrossRef]

Article

The Effects of Reduction and Thermal Treatment on the Recrystallization and Crystallographic Texture Evolution of 5182 Aluminum Alloy

Sofia Papadopoulou [1,2,*], Athina Kontopoulou [2], Evangelos Gavalas [1,2] and Spyros Papaefthymiou [2]

[1] ELKEME S.A., 61st km Athens-Lamia Nat. Road, 32011 Oinofyta, Viotia, Greece; egavalas@elkeme.vionet.gr

[2] Laboratory of Physical Metallurgy, School of Mining & Metallurgical Engineering, Division of Metallurgy and Materials, National Technical University of Athens, 9, Her. Polytechniou str., Zografos, 15780 Athens, Greece; ath.kontopoulou@gmail.com (A.K.); spapaef@metal.ntua.gr (S.P.)

* Correspondence: spapadopoulou@elkeme.vionet.gr; Tel.: +30-2262-60-4309

Received: 17 September 2020; Accepted: 15 October 2020; Published: 16 October 2020

Abstract: During forming, thickness reduction and thermal treatment affect the recrystallization and evolution of the crystallographic texture of metallic materials. The present study focuses on the consequences of rolling reduction of a widespread aluminum alloy with numerous automotive, marine and general-purpose applications, namely Al 5182. Emphasis is laid on the crystallographic texture and mechanical properties on both hot and cold-rolled semi-final products. In particular, a 2.8 mm-thick hot-rolled product was examined in the as-received condition, while two cold-rolled sheets, one 1.33 mm and the other 0.214 mm thick, both originating from the 2.8 mm material, were examined in both as-received and annealed (350 °C for 1 h) conditions. Electron back-scatter diffraction indicated the presence of a large percentage of random texture as well as a weak recrystallization texture for the hot-rolled product, whereas in the case of cold rolling the evolution of β-fiber texture was noted. In addition, tensile tests showed that both the anisotropy as well as the mechanical properties of the cold-rolled properties improved after annealing, being comparable to hot-rolled ones.

Keywords: DC Casting; 5182 alloy; crystallographic texture; hot rolling; cold rolling; crystallographic components; microstructural evolution; aluminum

1. Introduction

The need for lightweight, recyclable, malleable and strong materials has led to the use of aluminum alloys for a constantly increasing spectrum of industries with emphasis on the transportation sector both on land and sea serving automotive and marine applications, but also the food industry [1,2]. To gain tailored properties to match the needs of these industries various rolling and heat treatment processes must be fully controlled in order to tailor the microstructure, to enable an enhanced forming ability, and to gain expected final mechanical properties. In particular, the evolution of the crystallographic texture throughout forming and heat-treating stages, with regard to the final mechanical properties, is of great importance [3]. Liu et al. have examined the effect of direct casting (DC) versus continuous (CC) casting of Al 5052 exhibiting the effect of the casting process on the texture [4], while Yu et al. studied similar phenomena on Al 5182 [5]. Crystallographic texture is affected at every step of the thermomechanical procedure [6,7]. Therefore, the correlation between crucial process parameters (e.g., temperature, alloy composition, casting procedure, number of reduction steps etc.) and their influence to texture evolution are paramount for the optimization of the desired properties through the relevant steps.

The 5182 Al alloy (see Table 1) suits applications such as the beverage can lid as well as automotive structural parts and inner body parts [8–11]. Hot and cold rolling are used to match the final gauge. The more intense plastic deformation is applied to the alloy, the more the microstructure is affected. In the conventional rolling process, on the surface of the Al sheet that is produced, the shear stresses that are present lead to higher local strain. This results in more fine grains on the surface than in the middle of the produced sheet [12]. Concerning the effect of cold rolling on Al alloys, crystallographic components that consist the β fiber (brass, copper and S) are expected. Cold rolling is characterized by the activation of octahedral slip systems while hot rolling has the tendency to provoke the slip of non-octahedral systems. At the stacking fault energy is attributed the way that the plastic deformation will occur (slip and/or twinning). The Al alloys are plastically deformed through slip on the systems {111} <110> [13]. The texture components could be categorized in (i) rolling texture components (Copper, Dillamore or Taylor, S, Brass), (ii) shear texture components (rot-Cube$_{ND,}$ γ fiber, α fiber Inverse brass, E, F,) and (iii) recrystallization texture components (Cube, P, R and Goss) [14–19].

The amount of recrystallization components that will be formed during the production process depends on the prior deformed state. The aim is to obtain the desired equally balanced amount of texture components in order to achieve the best formability through the control of the earing behavior [17–19].

Two main theories had been widely discussed in the scientific community regarding texture evolution, the orientation of (i) the nucleation and (ii) the growth. Cube oriented grains are those that combine characteristics from the two theories altogether; they tend to grow from remained Cube bands and they tend to be favorable points for the growth of deformation crystallographic components [20]. The more intense effect on cube oriented grains is given by the rolling reduction percentage, the annealing conditions and the chemistry of the alloy [21]. It is crucial to decrease the presence of Cube component by controlling the hot rolling process, since a high amount of cube-oriented grains could lead to reduced drawability [16].

In general, cold rolling induces work hardening resulting in an increase of both yield strength and hardness. The application of heat-treatment (annealing), however, can reduce the extent of these effects, [14,15] and, in addition, it may reduce anisotropy. K. J. Kim et al. [16] studied the feasibility of deep drawing for aluminum alloys 5182 after annealing at 350 °C for different durations and concluded that the sheet, which had been annealed for 20 min exhibited the best drawability. The best combination of crystallographic components, according to the authors, to obtain the optimal drawability is γ fiber ND//<111> and rot-Cube$_{ND}$ {001}<110>. Due to the thermal loses during the insertion of the samples, however, other researchers considered a treatment time of one hour [22–26]. It has been documented that the rolling process results in the formation of a β fiber texture [27], consisting of the Cu{112}<111>, S{123}<634> and brass{110}<112> components. In addition, the application of heat treatment results in a further transformation of the grains into cube {001}<100>, P {011}<1-11>, Q {013}<3-31>, R {132}<4-21> and Goss {011}<100> components [14]. The crystallographic component R {124}<211> is similar to the crystallographic component S {123}<634>. Beck and Hu [28] pointed out that the R orientation can originate from the deformation texture by two different mechanisms. The first mechanism is the retained R from the rolling texture, where the stored dislocation energy is being reduced partially through extended recovery reactions, named as continuous recrystallization in the relevant literature. The second mechanism is about the forming of grains exhibiting R orientation genuinely via discontinuous recrystallization by the growth of nuclei into S oriented grains at the grain boundaries of the deformed bands and subsequent growth under the depletion of the rolling components, B, C and S. An R orientation is created through discontinuous recrystallization by the procedures of nucleation and growth, contrary to the display of a notable swiftness regarding the corresponding rolling texture, which is apparent to the ODFs with large scattering close to the R component. During discontinuous recrystallization the R component competes with the cube orientation with regard to nucleation. Since the R grains can grow up to substantial volume fractions due to the high amount of potential nucleation sites, they can outweigh the cube orientation [29].

Each texture component causes the material to form "ears" at different directions from RD (Cu, S and brass components results into the formation of 45°"ears" in deep drawn cups while the cube component strengthens the presence of 90°"ears"); hence to minimize earing, a proper combination of texture components should be obtained [17,29]. It is reported by O. Engler et al. [13] that transient orientations along the deformation process from the recrystallization towards the rolling texture—including the α-fiber orientations between Goss and Bs—provoke an asymmetry of the ears of the cups, where the 0° cited "ears" are of greater height than the 90°"ears". This asymmetry results in the formation of six ears, whereas in rare cases, two ears may appear. Hence, in order to achieve optimum properties for subsequent sheet-forming operations, a deep understanding of the evolution of texture during the production of the sheet is necessary [30]. In addition, the tensile properties of Al alloys are heavily influenced by the anisotropy, a parameter that is quantified through the r value [31]. Al alloys produced by means of cold rolling and annealing processes exhibit r values of the order of 0.55 to 0.85 [2]. The typical influences of rolling and recrystallization texture components on r values, as well as the {111}//ND components mentioned, are identified as relevant for increasing r values [26].

A number of models such as the Taylor/Bishop-Hill and Sachs are effectively utilized for the prediction of r values as well as earing behavior by taking into account texture data [32].

Nevertheless, the effect of the process stage that is examined in the present study on the texture versus mechanical properties (yield strength, tensile strength, elongation at fracture and R-values) has not been properly quantified yet for the examined rolling reduction percentages. Thus, the formation of well-balanced crystallographic texture leading to the desired properties is still a challenge for researchers [33].

The present study focuses on the effects of reduction (rolling) and heat treatment on the resulting crystallographic texture. Electron back-scatter diffraction (EBSD) and tensile tests are used to correlate microstructure and properties with the forming sequence and the thermal treatment background.

2. Materials and Methods

Three products were examined (Tables 1 and 2). Their details are as follows:

- A 2.8 mm thick, hot-rolled (h.r.) product (examined in the present paper in as-received state—as-rolled condition);
- Two cold-rolled (c.r.) sheets, a 1.3 mm (c.r.1) produced from h.r. and a 0.214 mm thick sheet (c.r.2) produced from c.r.1 without intermediate annealing (examined in the present paper in the as-received state and after laboratory annealing).

The 2.8mm thick hot-rolled product was examined in the as-received condition, whereas the 1.33 and 0.214 mm thick products were examined in both the as-received (cold-rolled) and heat-treated condition (annealed at 350 °C for 1 h). Annealing was conducted in the laboratory furnace.

Table 1. Description of examined samples.

Sample ID	Thickness (mm)	Condition
h.r.	2.8	Hot-rolled
c.r.1	1.3	Cold-rolled with 53% rolling reduction
c.r.a. 1	1.3	Cold-rolled with 53% rolling reduction and annealed
c.r.2	0.214	Cold-rolled 92% rolling reduction
c.r.a.2	0.214	Cold-rolled 92% rolling reduction and annealed

Table 2. Chemical composition of direct casted (DC) Al 5182 aluminum alloy.

Element	Si	Fe	Cu	Mn	Mg	Cr	Al
Wt%	0.11	0.21	0.04	0.35	4.60	0.01	Bal.

2.1. Metallographic Examination—Texture

Prior to the metallographic examination, all specimens were cut, in parallel to the rolling direction, before being cold mounted in resin. Specimens were afterwards grinded and polished by use of SiC papers as well as diamonds and silica suspensions. Metallographic examination was initially conducted by means of an optical microscope at a magnification range of 500×–1000× in order for the microstructure to be revealed and identified in the as polished condition. Through the observation of the samples in the optical microscope, an evaluation of the microstructure was undertaken in terms of a more general display of its characteristics (degree of recrystallization, direction of grains, etc.).Barker's electrolytic etch was afterwards applied in order for specimens to be examined in the etched condition as well according to references [33,34].

Through the utilization of a scanning electron microscope (SEM) FEI XL40 SFEG (current Thermo Fisher Scientific, Waltham, MA, USA), the microstructure of the samples was observed and qualitative elemental analysis of the particles at high magnification and the existence of secondary phases and measurement of average grain size were undertaken. Images of secondary electrons with a applied voltage of 20 kV at a working distance of 10 mm were obtained. The EDS technique was used to obtain atomic analysis spectra of some observed intermetallic phases.

The electron back-scatter diffraction (EBSD scans were collected employing a high-speed Hikari EDAX camera, EDEN Instruments SAS, Valence Cedex, France) electron diffraction microscopy technique provides quantitative data on the structure and crystallographic nature of compatible test specimens, making a significant contribution to the characterization of a microstructure. Specifically, EBSD determines the size and orientation of the grains, the crystallographic tissue, the nature of the grain boundaries, the phase identification in multiphase materials, as well as the distinction of submicroscopic structures. The above data are used to understand the relationships between microstructure characteristics and macroscopic properties so they are used for alloy design, material characterization and failure analysis.

For the EBSD analysis, the sample is placed inside the SEM in the as polished condition with a certain inclination of 70° relative to the incident electron beam. Voltage was set at 20 kV, while the step size was set at 0.1 μm to 1 μm depending on the material condition and the magnification utilized (100×–2000×) (examined areas 30 × 30 μm^2–150 × 150 μm^2). Working distance was set at 10 mm. Scanning data was then further analyzed by means of a post - processing software package in order for the various texture components to be identified and calculated.

2.2. Tensile Testing

Testing was performed on a tensile machine with a capacity of up to 30 kN as per the guidelines of ASTM E 646-00 [35] and ASTM E 517-00 [36] with parallel use of static extension meter according to the ISO 6892-1 standard at room temperature. Rolled lines were easily visible on the surface of the samples. The tensile specimens were produced by hydraulic pressing according to the specifications of the tensile test. Along the reduced cross-section, the specimens were characterized by a thickness of 0.4 mm, a width of 12.55 mm and a total length of 158 mm. In total, 3 tensile specimens were created to ensure repeatability. The bending specimens were cut with a cutting tool for thin rolled products. The specimens had dimensions of 2.5 × 5 mm^2. A total of 45 specimens were produced [3 samples for each testing direction (0°, 45°, 90°) for each examined sample]. The standard deviation of all measurements was <1.5. In addition to the $Rp_{0.2}$ yield strength, the tensile strength and the elongation at fracture, the anisotropy parameter r was also measured in all directions.

As for the anisotropy parameter r, the following are calculated for the three directions (0°, 45°, 90°), relative to the rolling direction according to Equation (1):

$$r = \frac{ln\frac{w_1}{w_0}}{ln\frac{t_1}{t_0}} \tag{1}$$

with w_0 and t_0 being the width and thickness of the tensile specimen in the gauge region prior testing, respectively. In addition, w_1 and t_1 are the width and thickness of the specimen in the same region just before necking occurs. It is noted that tensile testing speed was set at 10 mm/min throughout the test.

3. Results

The grain structure was revealed after Barker's etching. The hot rolled sample exhibited equiaxed grain structure. After the first cold roll pass, grains became elongated. After annealing, the grains retrieved their equiaxed morphology, which was retained even after the final cold roll pass. The final product showed a fully recrystallized microstructure (Figure 1).

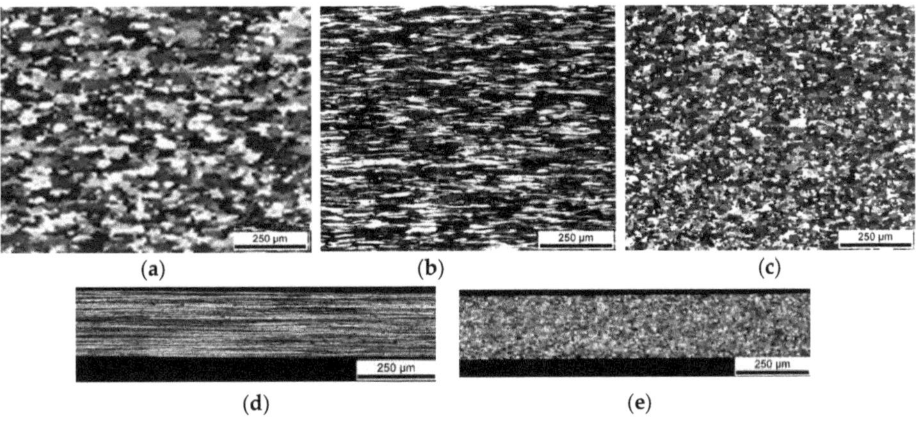

Figure 1. The microstructure of (**a**) hot-rolled sheet, (**b**) cold-rolled 1.3 mm sheet, (**c**) cold-rolled 1.3 mm annealed sheet, (**d**) cold-rolled 0.214 mm sheet and (**e**) cold-rolled 0.214 mm annealed sheet after Barker's etching.

Second phase particles of $Al_6(Mn, Fe)$ and Mg_2Si are dispersed all over the matrix of the examined samples. Indicative SEM micrographs of the particle dispersion in the matrix of h.r. and c.r.2 samples are shown in Figure 2.

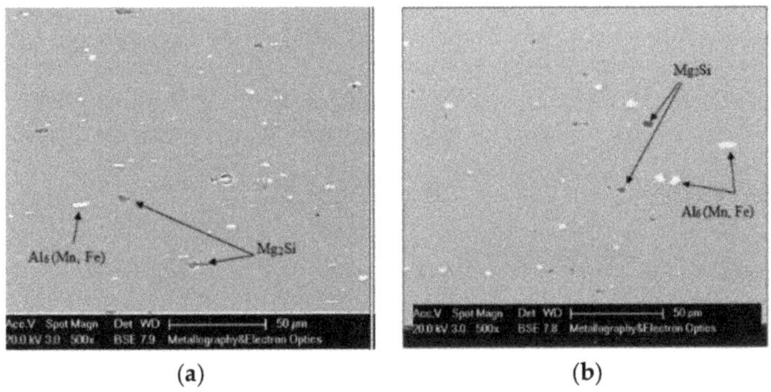

Figure 2. Scanning electron microscopy (SEM) images showing (**a**) h.r. sample and (**b**) c.r.2.

Intensities from inverse pole figure (IPF) and pole figure (PF) diagrams are related with the texture intensity of the examined sample. Red color indicates the points, where the highest texture intensity is observed. The lower intensity indicates higher isotropy of the texture and, consequently, the higher isotropy of mechanical properties as well. PF and IPF diagrams exhibit the same information from a different aspect. The findings originating from EBSD data are graphically illustrated in Figure 3:

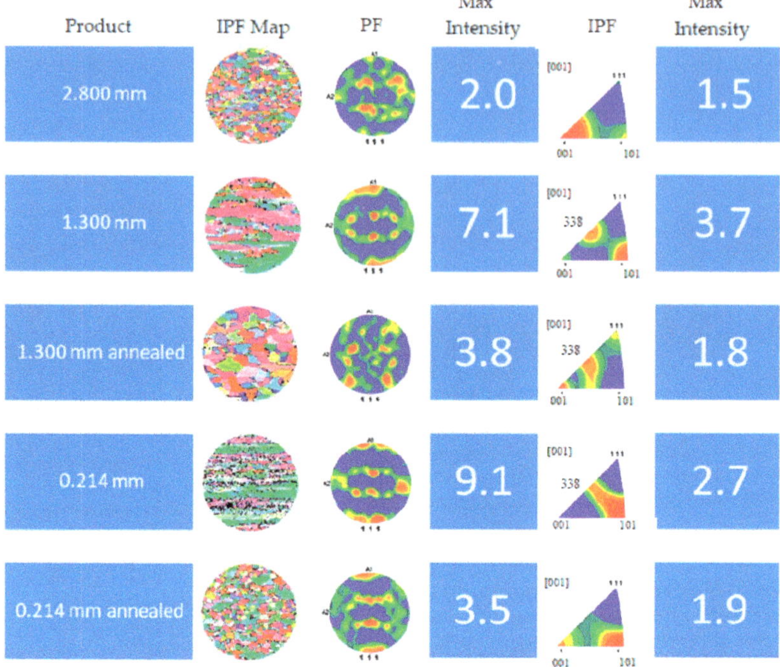

Figure 3. Electron back-scatter diffraction (EBSD) data for the examined products.

Figure 3 presents the IPF maps and microstructures as well as the PF for each product examined. The relevant intensities are also presented on the right side of Figure 3. The h.r. sample is partially recrystallized showing a typical hot-rolled PF exhibiting relatively low texture intensity (2.0 in the PF diagram and 1.5 in the IPF diagram). After 53% thickness reduction, sample c.r.1 grains tend to be more elongated, while the IPF diagram indicates a shift towards {111} orientation with measured intensity 7.1 and 3.7 for PF and IPF diagrams respectively. The annealing creates a fully recrystallized and equiaxed grain structure, while the intensity is low (3.8 and 1.8 in the PF and the IPF diagram respectively). At the same time, a crucial difference is observed from the h.r. to the c.r.a.1 sample: a weak {111} and a strong {338} component is formed. After the final cold pass, the higher amount of texture intensity has been measured (9.1 in the PF and 2.7 in the IPF diagram respectively) (Figures 3 and 4). After the final annealing, the texture intensity reaches similar values with the ones measured on the c.r.a.1 sample (3.5 and 1.9 in the PF and in the IPF diagram respectively) (see Figure 3). At c.r.a.2 the {001} had be reformed after been seen in h.r. and in c.r.a.1.

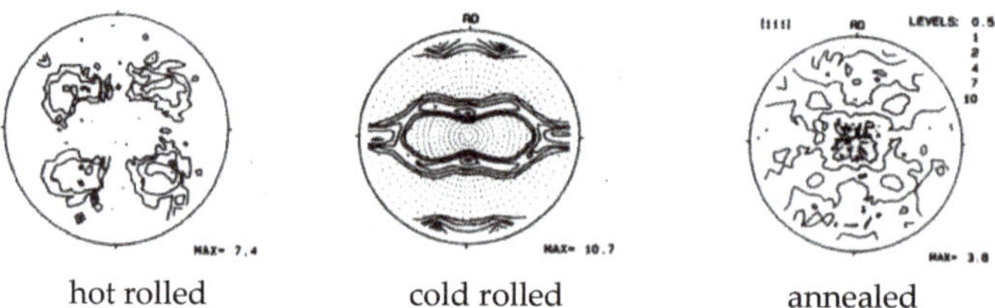

Figure 4. Typical textures [(111) pole figures] in different processing stages of Al-alloy sheet [32].

For reference, the typical appearance of PF in relation to the rolling process and heat treatment applied are given in Figure 4 [37].

The Al sheet texture exhibits three main categories of component, recrystallization (cube, P, R), rolling (brass, copper, Goss, S, Dilamore or Taylor) and shear (rotated cube, inverse brass, E, F) texture components. The h.r. sample, mostly exhibited recrystallization texture accompanied by a significant amount of rolling texture (Figure 5). The main components detected in this state are Q, R and brass. After 52.5% thickness reduction, the rolling texture increased and the main components detected are R and Goss from the remained texture recrystallization and intense S, brass, copper and Taylor.

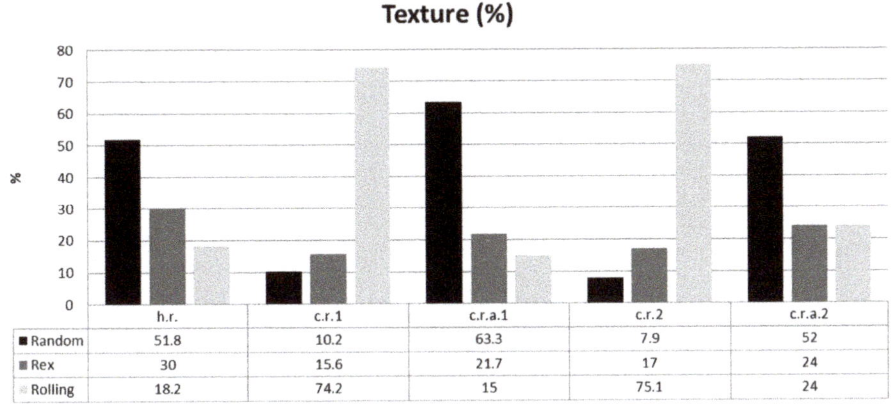

	h.r.	c.r.1	c.r.a.1	c.r.2	c.r.a.2
Random	51.8	10.2	63.3	7.9	52
Rex	30	15.6	21.7	17	24
Rolling	18.2	74.2	15	75.1	24

Figure 5. Texture percentage chart.

The effect of annealing changed the components analogies increasing the random texture. Q, R and Goss increased leaving just a small amount of S and Taylor from the previous deformation. The final pass of 92% thickness reduction led to the formation of high amounts of S and brass, followed by Taylor and Copper. The representative recrystallization component is R. The ultimate annealing process provoked the formation of a high amount of R coexisting with brass and S (see Figure 5). Cold rolling of the h.r. sample, led to an increased amount of both rolling and recrystallization texture percentage for the c.r.1 and c.r.2 samples. It also led to increased anisotropy (see the increased intensities PF and IPF diagrams). Annealing treatment formed low PF and IPF intensities, which resemble to the original h.r. sample observed intensities.

The random texture of both annealed sheet samples increased significantly reinforcing the isotropic behavior of these materials with approximately 1% normal anisotropy. This could be explained by the fact that in c.r.a.1 the nucleation of the cube component was inadequate (r-CubeRD) and the amount of recrystallization texture was higher than the remaining deformation texture. In contrast the c.r.a.2 exhibited greater volume fractions of the cube and R components, which favor the earing formation of 0/90° ears. In both cases the recrystallization texture in the annealed samples is higher than the cold-rolled samples before.

EBSD findings are analytically presented in the following chart, where the exact percentages of the random, recrystallization (Rex) and rolling texture are shown for each examined case (Figure 5).

The mechanical properties after annealing for both cold-rolled products exhibited significant similarity with regard to the h.r. mechanical properties. Tensile testing results are given in Figures 6–8. In particular h.r. exhibited yield strength of 139 MPa at 0°, of 144 at 45° and of 145.5 at 90° whereas c.r.a.1 showed 155 MPa, 148.5 MPa and 154.5 MPa and c.r.a.2 161 MPa, 152.5 MPa and 160.5 MPa, respectively. Concerning the tensile strength, h.r. exhibited 301.5 MPa at 0°, 298 MPa at 45° and 299.5 MPa at 90° while c.r.a.1 showed 312.5 MPa, 303 MPa and 301.5 MPa and c.r.a.311.5 MPa, 298.5 MPa and 303 MPa, respectively. The measured values of the h.r, c.r.a.1 and c.r.a.2 confirmed that regarding the yield strength and the tensile strength, the annealing temperature restored the mechanical properties after the deformation, but did not have such an intense impact on crystallographic components (see Figure 3).

The PF intensity was selected as an indicator of intense texture presence and, therefore, less isotropic behavior.

Yield strength and tensile strength increased to an analogous extent in all directions after cold rolling and decreases after annealing (Figures 6 and 7). The 0° direction exhibited slightly higher strength values compared to the other directions for all but the 2.8 mm thick products.

In both annealed cases (c.r.a.1 and c.r.a.2) the elongation at fracture was increased which can be related to the reduction of dislocation density and the new formed grain size and morphology (Figure 8).

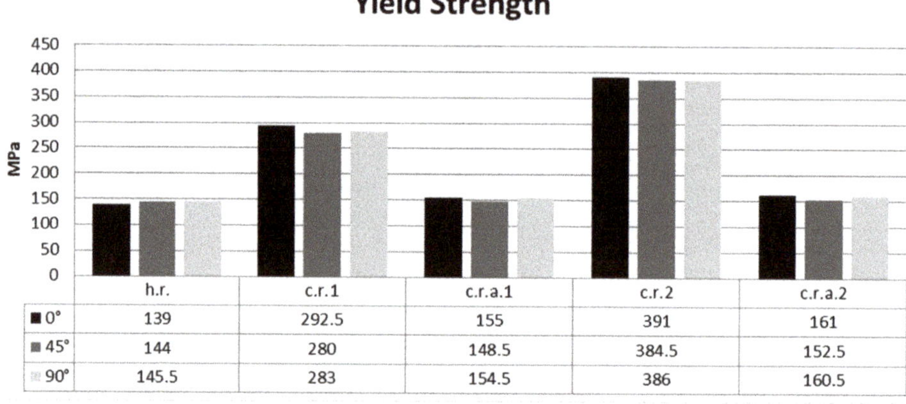

Figure 6. Yield strength data for all products and conditions.

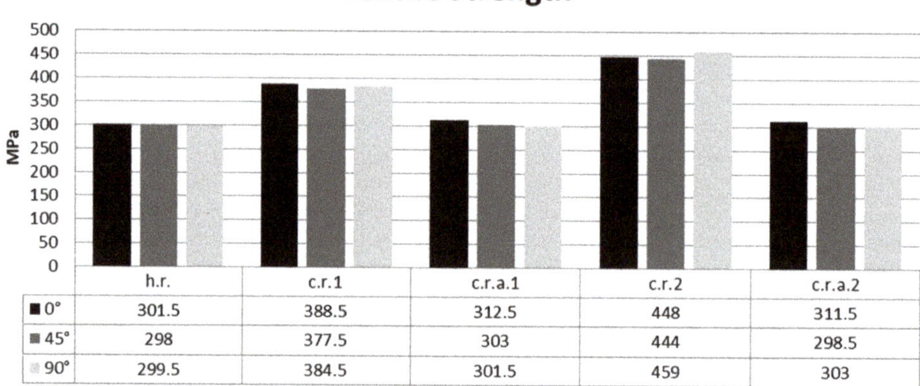

Figure 7. Tensile strength data for all products and conditions.

Elongation at fracture

	h.r.	c.r.1	c.r.a.1	c.r.2	c.r.a.2
■ 0°	23.5	11.5	22	6	20
■ 45°	26.5	14	25.5	7.5	23.5
▩ 90°	25	14	22.5	10	26

Figure 8. Elongation at fracture data for all products and conditions.

On a similar note, the r value was also influenced as a result of the various processing steps. In particular, cold rolling resulted into a reduction of the 0° r value in the rolling direction, whereas a significant increase was noted in the 45° and 90° directions. Similar observations were made for the 1.33 and 0.214 mm thick cold-rolled products. Annealing resulted in an increase of the r value in the rolling direction, whereas the r values in the 45° and 90° directions decreased. As a result, all r values after annealing closely resembled the relevant values of the original h.r. sample (see Figure 9). Sample c.r.1 after 53% thickness reduction, exhibited decreased elongation (<14%). After the annealing process, elongation increased to >22%, which dropped to <10% after 92% reduction. After the final annealing, elongation reached along all the 0°, 45° and 90° directions, respectively >20%.

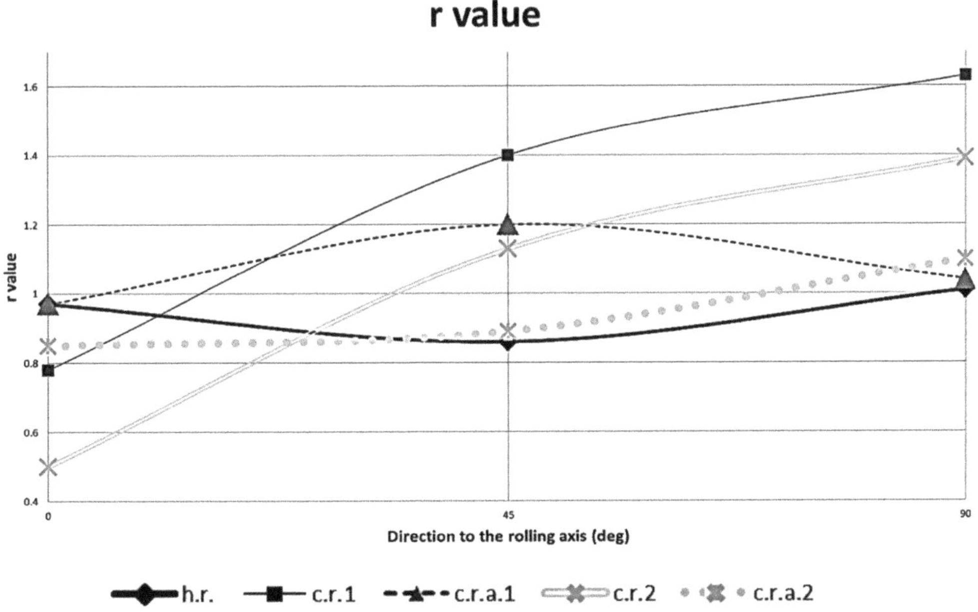

Figure 9. The r average in relation to direction (0°, 45°, 90°).

4. Discussion

The hot-rolled sheet exhibited a balanced amount of rolling and recrystallization texture components along with random texture. Cold rolling of the 2.8 mm hot-rolled sheet led to an increased amount of rolling texture, at the expense of both random and recrystallization textures. The effect of annealing was evident through the reduction of the rolling and recrystallization components percentages, whereas the random texture percentage increased. The amount of rolling texture components decreased whereas the amount of recrystallization and random components increased.

Cold rolling of the original 2.8 mm thick sheet resulted in increased anisotropy as indicated by the increased intensities PF and IPF diagrams. In contrast, post-cold-rolling annealing resulted in low PF and IPF intensities indicating a behavior relating to a more isotropic material, comparable to the original 2.8 mm thick hot-rolled product. This is beneficial due to the increase of fracture elongation and the reduction of anisotropy and elongated grain morphology, which can assume increased formability. In addition, cold rolling resulted in increased yield and tensile strength values as well as a significant decrease of fracture elongation due to the presence of fibrous grain with increased anisotropy. The dislocation density and grains size and morphology have major effects on elongation at fracture of the tensile specimens. PF intensity controls anisotropy of those mechanical properties, but hardly their overall level.

This effect was more pronounced in the 0.214 mm thick product that had to undergo two reductions instead of a single one, as was the case for the 1.33 mm thick product. Post cold rolling annealing restored the mechanical properties of the two cold-rolled products in levels comparable to the 2.8 mm thick hot-rolled product and as a result elongation and formability were improved. Cold rolling texture is transformed to the cube component during annealing, which favors its performance with regards to bending, whereas the addition of {111}<uvw> components such as {111}<110> during annealing favors deep drawability.

The random texture of both the annealed sheet samples was increased significantly which could be understood due to the suppression of nucleation and subsequent growth of the cube orientation which is expected to be strong in recrystallized aluminum rolled products. The almost homogeneous r value of the h.r., c.r.a.1 and c.r.a.2, a better formability could be expected [37].

Suitable texture control relating to specific applications is required for the improvement of sheet metal behavior with regards to further forming operations (e.g., bending and deep drawing).

Structural variables could be used in measurements of dislocation cell morphology and material behavior. Therefore, the dependence of the cell morphology should be identified from the thermomechanical processes upon crystallite lattice orientation. Nevertheless the scan of areas of the order of a few μm or mm requires small step sizes in the case of highly deformed samples that exhibit good statistical accuracy, however, according to relevant literature [4–8]. As such, data obtained from automated EBSD analysis of highly deformed microstructures can lead to a deeper understanding of the deformation behavior and the recrystallization and microstructure evolution sequences with regards to the applied deformation steps. Thus, the final microstructure and mechanical properties can be tailored accordingly with regard to the final thickness and the required strength levels as well as the necessary subsequent formability, which is the ultimate case.

EBSD data showed that after the first cold roll pass (c.r.1), crystallographic texture mainly contains β fiber texture (which was present even from the hot rolling process) with strong Taylor component. After annealing treatment, texture evolved mainly to Q, R and weak β fiber texture with the Taylor component maintaining the highest intensity. The second and final pass evolved the observed texture into a mixture of β fiber texture with the S component having the highest intensity and simultaneously strong Goss component; no significant cube texture was found. A possible path between directions {101} and {338}, could be a way for the nucleation of cube crystallographic components. At c.r.a.2 the {001} had be reformed after being seen in h.r. and in c.r.a.1.

Recrystallization exhibits characteristics of phase transformation in which the replacement of deformed material due to the creation of nuclei and growth of recrystallized grains can lead to significant changes of the resulting texture. The latter does not lead to precise orientation relationships among deformed and recrystallized material, contrary to the phase transformations. It should be stated that recrystallization does not always lead to texture transformations, especially when large volume fractions of second phase particles are present or when the deformation was axisymmetric.

The quantitative correlation between PF intensity and mechanical properties could be attributed to the anisotropy increase that takes place after each mechanical process, but it is important to observe the analogy of these values towards texture. The impact of texture components is strong on the variations of r-value for different directions. In all examined cases, the texture component occupies >92% of the total volume fraction of grains. The remaining volume fraction is populated by randomly oriented grains. The r-value is non-uniform for all the examined direction for each sample. It is clearly seen from Figure 9 that the optimal texture for the smallest possible in-plane anisotropy clearly gives an average r-value less than 1.0 (exhibited by h.r.).

The r values of the Al sheet, when measured after hot rolling, can be regarded as more appropriate amongst those of all measured samples. After annealing of the 92% reduced sheet, r measurements indicated a quite good forming behavior, but the effect of the final pass towards the anisotropy and r values at 0°, 45°and 90°could not be eliminated throughout annealing. A higher, than expected for this Al alloys, r value was observed at 90°, a behavior that can be attributed to the higher amount of rolling texture and especially to the presence of the S component. Therefore, a reduced amount of thickness reduction, the number of rolling passes and the lubrication could provoke better results and, possibly, the ideal percentage of rolling and annealing crystallographic components.

5. Conclusions

This work demonstrates that the mechanical properties of Al5182 can be influenced by the deformation sequence as well as the subsequent annealing conditions. Moreover, texture is especially important with regards to formability, especially for the deep drawing of thin sheets.

More specifically:

- After the application of the various rolling steps as well as the appropriate heat treatment (350 °C for 1 h), even in the material that was submitted to 92% thickness reduction (c.r.2-0.214 mm), sheets exhibited an increase of yield and tensile strength, mechanical properties closely resemble those of the reference material (as-received condition after hot rolling, h.r.). The formation of {110}//RD texture components in h.r. favors the increase of the elongation of the alloy. In addition the relevant PF and IPF intensities increased after thickness reduction as a result of increased anisotropy. This is also evident by the recorded r values, which are heavily influenced by cold rolling whereas after the annealing process they are restored to the as-received product levels. The same applies for the tensile properties and the intensities of the PF and IPF diagrams.
- The initial texture (formed from hot rolling) has a strong influence on the distribution of component intensities along the β fiber. The β-fiber maximum intensity was lower than 10, revealing a low yet remaining amount of rolling texture in the material after the intermediate annealing. Equivalence among rolling and recrystallization crystallographic components is essential to retain formability and toughness in order to form a fine and equiaxed grain structure.
- To increase the r value towards all directions, R orientation [123]<634> is expected to be most efficient, but this orientation shows a low r value of about 0.8 mainly at the 0° and 45° directions. Therefore, the Goss orientation [011]<100>, which also increases the r value must be developed to some extent. [001]<uvw> textures such as the cube orientation are non-preferable with regard to the improvement of the r value. The variations observed in the products that have been annealed were coherent with the R texture component. The r-value of the as-rolled material was consistent with the presence of cold-rolling texture components.

Author Contributions: Conceptualization, E.G., S.P. (Spyros Papaefthymiou); Methodology, E.G., S.P. (Spyros Papaefthymiou), S.P. (Sofia Papadopoulou); Software, S.P. (Sofia Papadopoulou) and A.K.; Validation, S.P. (Sofia Papadopoulou); Formal Analysis, S.P. (Sofia Papadopoulou), E.G. and A.K.; Investigation, A.K. and S.P. (Sofia Papadopoulou); Resources, S.P. (Spyros Papaefthymiou); Data Curation, S.P. (Sofia Papadopoulou) and A.K.; Writing—Original Draft Preparation, S.P. (Sofia Papadopoulou) and A.K.; Writing—Review & Editing, E.G. and S.P. (Spyros Papaefthymiou); Visualization, S.P. (Sofia Papadopoulou); Supervision, E.G. and S.P. (Spyros Papaefthymiou); Project Administration, S.P. (Sofia Papadopoulou). All authors have read and agreed to the published version of the manuscript.

Funding: This research received no external funding.

Acknowledgments: The authors would like to express their gratitude to ELKEME SA and ELVAL SA managements for all the kind support. Pantazopoulos, Vazdirvanidis are highly appreciated for the fruitful discussions and Toulfatzis for the mechanical testing.

Conflicts of Interest: The authors declare no conflict of interest. The funders had no role in the design of the study; in the collection, analyses, or interpretation of data; in the writing of the manuscript; or in the decision to publish the results.

References

1. Hatch, E. *Aluminum: Properties and Physical Metallurgy*; ASM International: Almere, The Netherlands, 1984; pp. 134–199. [CrossRef]
2. Cantor, B.; Grant, P.; Johnston, C. *Automotive Engineering: Lightweight, Functional, and Novel Materials*; Taylor and Francis: Abingdon, UK, 2008.
3. Bunge, H.J. *Texture Analysis in Materials Science*; Butterworhs: London, UK, 1982.
4. Liu, J.; Morris, J.G. Recrystallization microstructures and textures in AA 5052 continuous cast and direct chill cast aluminum alloy. *Mater. Sci. Eng. A* **2005**, *385*, 453–454. [CrossRef]
5. Yu, X.F.; Zhao, Y.M.; Wen, X.Y.; Zhai, T. A study of mechanical isotropy of continuous cast and direct chill cast AA5182 Al alloys. *Mater. Sci. Eng. A* **2005**, *394*, 376–384. [CrossRef]
6. Bunk, W.; Lucke, K.; Masing, G. Rolling and recrystallization textures of aluminum. *Z Metallkd* **1954**, *45*, 584–593.
7. Lücke, K. Rolling and recrystallization texture of aluminum. Texture of cold-rolled aluminum. *Z Metallkd* **1954**, *45*, 86–92.
8. Romhanji, E.; Popovic, M.; Glisic, D.; Stefanovic, M.; Milovanovic, M. On the Al-Mg alloy sheet for automotive applications: Problems and solutions. *Metallurgija* **2004**, *10*, 202–216.
9. Davis, J.R. *Alloying: Understanding the Basics*; ASM International: Novelity, OH, USA, 2001.
10. Hirsch, J. Texture and anisotropy in industrial applications of aluminium alloys. *Arch. Metall. Mater.* **2005**, *50*, 21–34.
11. Miller, W.S.; Zhuang, L.; Bottema, J.; Wittebrood, A.J.; De Smet, P.; Haszler, A.; Vieregge, A. Recent development in aluminium alloys for theautomotive industry. *Mater. Sci. Eng. A* **2000**, *280*, 37–49. [CrossRef]
12. Alharthi, N. Microstructure Evolution of Asymmetrically Rolled AA-5182. In Proceedings of the ASME District A 2011 Student Professional Development Conference, Philadelphia, PA, USA, 2 June 2011.
13. Bate, P.; Oscarsson, A. Deformation banding and texture in hot rolled Al–1.0Mn–1.2Mg alloy. *Mater. Sci. Technol.* **1990**, *6*, 520–527. [CrossRef]
14. Humphreys, F.J.; Hatherly, M. *Recrystallization and Related Annealing Phenomena*; Elsevier: Amsterdam, The Netherlands, 2004.
15. Hirsch, J.; Al-Samman, T. Superior light metals by texture engineering: Optimized aluminum and magnesium alloys for automotive applications. *Acta Mater.* **2013**, *61*, 818–843. [CrossRef]
16. Kim, K.J.; Won, S.-T.; Park, J.-H. Texture analysis of 5182 aluminum alloy sheets for improved drawability by rolling process. *Mater. Werkst.* **2012**, *43*, 373–378. [CrossRef]
17. Theis, H. *Handbook of Metalforming Processes*; CRC Press: New York, NY, USA, 1999.
18. Zhang, L.; Wang, Y.; Yang, X.; Li, K.; Ni, S.; Du, Y.; Song, M. Texture, microstructure and mechanical properties of 6111 aluminum alloy subject to rolling deformation. *Mater. Res.* **2017**, *20*, 1360–1368. [CrossRef]
19. Kocks, U.F.; Tome, C.N.; Wenk, H.R. *Texture and Anisotropy*; Cambridge University Press: Cambridge, UK, 2000.
20. Engler, O.; Vatne, H.E.; Nes, E. The roles of oriented nucleation and oriented growth on recrystallization textures in commercial purity aluminium. *Mater. Sci. Eng. A* **1996**, *205*, 187–198. [CrossRef]

21. Bennett, T.A.; Sidor, J.; Petrov, R.H.; Kestens, L.A. The effect of intermediate annealing on texture banding in aluminum alloy 6016. *Adv. Eng. Mater.* **2010**, *12*, 1018–1023. [CrossRef]
22. Totten, G.E.; Funatani, K.; Xie, L. *Handbook of Metallurgical Process Design*; Marcel Dekker Inc.: New York, NY, USA, 2004.
23. Kim, H.-W.; Lim, C.-Y. Annealing of flexible-rolled Al–5.5wt%Mg alloy sheets for auto body application. *Mater. Des.* **2010**, *31*, S71–S75. [CrossRef]
24. Okada, M.; Hirano, S. Effect of Annealing Condition on Earing and Texture Formation in Cold Rolled AA5182 Aluminum Alloy. In Proceedings of the 13th International Conference on Aluminum Alloys (ICAA13), Pittsburgh, PA, USA, 25 June 2012.
25. Liu, W.C.; Zhai, T.; Man, C.-S.; Morris, J.G. Quantification of recrystallization texture evolution in cold rolled AA 5182 aluminum alloy. *Scr. Mater.* **2003**, *49*, 539–545. [CrossRef]
26. Liu, W.C.; Morris, J. Effect of initial texture on the recrystallization texture of cold rolled AA 5182 aluminum alloy. *Mater. Sci. Eng. A* **2005**, *402*, 215–227. [CrossRef]
27. Leffers, T.; Ray, R. The brass-type texture and its deviation from the copper-type texture. *Prog. Mater. Sci.* **2009**, *54*, 351–396. [CrossRef]
28. Barrett, C.S. *The Structure of Metals*, 2nd ed.; McGraw-Hill: New York, NY, USA, 1952.
29. Hutchinson, W.B.; Ekström, H.E. Control of annealing texture earing in non-hardenable aluminum alloys. *Mater. Sci. Technol.* **1990**, *6*, 1103–1112. [CrossRef]
30. Engler, O. Simulation of rolling and recrystallization textures in aluminium alloy sheets. *Mater. Sci. Forum* **2007**, *550*, 23–34. [CrossRef]
31. Savoie, J.; Jonas, J.J.; MacEwen, S.R.; Perrin, R. Evolution of r-value during the tensile deformation of aluminium. *Textures Microstruct.* **1995**, *23*, 149–171. [CrossRef]
32. Jamaati, R.; Toroghinejad, M.R. Effect of alloy composition, stacking fault energy, second phase particles, initial thickness, and measurement position on deformation texture development of nanostructured FCC materials fabricated via accumulative roll bonding process. *Mater. Sci. Eng. A* **2014**, *598*, 77–97. [CrossRef]
33. Voort, G.F.V. *Metallography, Principles and Practice*; McGraw-Hill: New York, NY, USA, 1984.
34. ASTM. *E407-07 (2015) Standard Practice for Microetching Metals and Alloys*; ASTM International: West Conshohocken, PA, USA, 2015.
35. ASTM. *E646-00: Standard Test Method for Tensile Strain-Hardening Exponents (n-Values) of Metallic Sheet Materials*; ASTM International: West Conshohocken, PA, USA, 2000.
36. ASTM. *E517-00: Standard Test Method for Plastic Strain Ratio r for Sheet Metal*; ASTM International: West Conshohocken, PA, USA, 2000.
37. Hirsch, J.R. Aluminium alloys for automotive application. *Mater. Sci. Forum* **1997**, *142*, 33–50. [CrossRef]

Publisher's Note: MDPI stays neutral with regard to jurisdictional claims in published maps and institutional affiliations.

Article

Thermal Camber and Temperature Evolution on Work Roll during Aluminum Hot Rolling

Evangelos Gavalas [1,2,*] and Spyros Papaefthymiou [2]

[1] ELKEME S.A., 61st km Athens-Lamia Nat. Road, Oinofyta, 32011 Viotia, Greece
[2] Laboratory of Physical Metallurgy, Division of Metallurgy and Materials, School of Mining & Metallurgical
     Engineering, 9, National Technical University of Athens, Her. Polytechniou str., Zografos,
     15780 Athens, Greece; spapaef@metal.ntua.gr
* Correspondence: egavalas@elkeme.vionet.gr

Received: 15 September 2020; Accepted: 28 October 2020; Published: 28 October 2020

Abstract: Flatness is an important quality characteristic for rolled products. Modern hot rolling mills are equipped with actuators that can modify the uneven thickness distribution across the width of the strip (crown), taking into account online measurements of various process parameters such as temperature, force and exit strip profile, either automatically or manually by the operator. However, the crown is also influenced by many parameters that cannot easily be measured during production, such as work roll temperature evolution through thickness and roll geometric variation due to thermal expansion (thermal camber). These have an impact on the strip flatness. In this paper, a thermo-mechanical finite element model on LS-DYNA™ software was utilized to predict the influence of process parameters, and more specifically strip temperature, cooling strategy (application of cooling on the entry or entry and exit side simultaneously) and roll core temperature, on the evolution of roll temperature and thermal camber. The model was initially validated with industrial data. The results indicate that the application of both entry and exit cooling is ~30% more efficient compared to the entry cooling only, thus the thermal camber will be reduced by 2 μm. A hotter roll (380 K) is more stable compared to the cold roll (340 K), showing also an improvement of 2 μm. The hotter roll will also reach a thermal steady state on the surface faster compared to the colder one, without making a significant difference on the steady state temperature. Strip temperature plays a roll in the thermal camber evolution, but it is a less important parameter compared to cooling strategy and roll temperature.

Keywords: aluminum; hot rolling; LS-DYNA; crown; camber; cooling; temperature evolution

1. Introduction

Aluminum alloys are constantly increasing in terms of their consumption and spectrum of industrial use, with emphasis, amongst others, on the transportation sector, the food industry and construction. A large amount of the materials used in these sectors are strips/sheets that are formed or cut to the final shape using special blanks with very tight tolerances. Such applications require very large amounts of material with strict quality characteristics. One of the most important quality characteristics is the geometric accuracy. The increased requirements for rolled aluminum products with high geometric accuracy and high productivity have triggered many studies related to the effect of the rolling parameters on the strip's profile.

An important geometric requirement is crown, wherein the thickness varies through the width of the strip. Crown is associated with the roll deflection due to the separating forces developed between the strip and the work rolls, causing the deformed rolls to not create a perfectly rectangular gap, but one that is curved from the top and bottom side (work roll sides) and as a result defining the geometry of the strip [1–7].

Conventional 4-high rolling mills are equipped with hydraulic systems that allow the bending of the rolls in order to adjust the strip's crown to the required profile. In the production environment, the bending set point is predetermined and defined during the pass schedule design [1]. Many researchers have utilized analytical and numerical modeling techniques in the effort to predict the work roll bending [1–7]. Guo (1989) [2] implemented a simple model calculating the work roll bending for each individual schedule. The results were used to classify pass schedules with similar crowns. Sikdarl et al. [3] provided the bending set point for crown control by modeling the roll stack deflection of a hot strip mill. Steinboeck et al. [4] applied nonlinear constitutive equations and a change of coordinates to obtain a time-free formulation to calculate the work roll bending. Fukushima et al. [5] developed an online analytical model, coupled with an FE model that calculated the rolling load distribution, to accurately predict the strip profile. This method enabled the rolling of substantially different products in the same production line. Wang et al. [6] calculated the crown by modeling the process in two steps. The first step was to calculate and predict the rolling force distribution and secondly to use the results as input data in order to calculate the roll deflection. Gavalas et al. [1] implemented an elastic-viscoplastic finite element analysis in order to predict the crown under various rolling conditions for pass schedule strategy allocation. Shigaki et al. [7] predicted roll stack deflection by coupling a commercial FEM model with a multi-slab model for strip deformation. Such approaches have shown good results and are adapted in the real industrial environment in order to either define set points or counteract the effects of variations induced in the process. When these variations become significant, then predictive models have reduced accuracy and the adaptive bending system sometimes either cannot completely correct the crown or it can introduce inhomogeneities in the material.

Work roll thermal camber has a direct influence on profile and is caused by the temperature increase and the presence of temperature gradient due to heat conducted from the hot strip to the roll. Radial thermal expansion differs across the roll width due to axial heat flux towards the roll sides. The result is a greater increase at the center compared to the edges, creating a thermal camber [8]. The thermal camber changes dynamically during production since the thermal state is influenced by many parameters, such as the roll's core temperature, the strip temperature, the length of the pass, inter-pass time, rolling time since the last work roll replacement, etc.

Many studies focus on the prediction of work roll temperature profile and thermal camber. Tseng et al. [9] developed an analytical solution that predicted the thermal expansion at the roll center plane, and estimated the crown by taking the edge diameter as reference. Sturmer et al. [10] investigated if the combination of models could be replaced with a fast enough 3-D model of the roll shape that can be used online during operation. Jiang et al. [11] derived an online model using differential equations in conjunction with a neural network model to predict the thermal crown in a hot rolling process. Many researchers have studied the thermal crown using the finite difference method (FDM), which is a common numerical method for multidimensional heat transfer problems. Ginzburg et al. [12] developed a simple but efficient model called Coolflex that could be applied in different mill configurations and would analyze the effect of rolling parameters on the thermal profile. Atack et al. [8] incorporated an FDM thermal camber model, whereby spray patterns were the independent variables to determine an optimization program. Lin et al. [13] considered a uniform heat source across the strip width and employed an FDM analysis to compute work roll temperature profile and thermal expansion. Abbaspour et al. [14] used FDM to solve the energy equation in the radial and axial direction, also taking into consideration strip width, time between the passes, strip temperature and thickness reduction. The computational cost of the previous methods is low, but many assumptions have to be considered in the model, limiting the accuracy of the calculation results.

The other recognized modeling method is based on the finite element method (FEM)m and has been widely exploited by many researchers for thermo-mechanical simulations due to the increased quality of the results and the possibility of applying realistic boundary conditions and constraints in complicated models. Guo et al. (2006) [15] analyzed the temperature field and thermal crown of the roll with a simplified finite element method (FEM). Benasciutti et al. [16] and Li et al. [17] proposed an FEM model to calculate the thermal stresses occurring in the work roll due to the non-uniform temperature distribution on the roll surface due to the hot rolling process. Trull et al. [18] developed an FEM model incorporating all major mill components and engineering and process conditions, including thermal camber, for the simulation of shape evolution. Bao et al. [19] investigated the temperature distribution of the work roll, integrating FEM coupled electromagnetic–thermal analysis in order to include the effect of roll induction heating on the surface temperature. Deng et al. [20] developed an analytical FEM model to investigate the thermal and oxidation behavior of an HSS (High Speed Steel) work roll during hot rolling.

In this paper, a thermo-mechanical finite element model on LS-DYNA™ software was designed to study the influence of process parameters, and more specifically strip temperature, cooling unit strategy (entry cooling and entry/exit cooling) and roll average temperature, on the evolution of roll temperature and thermal camber during aluminum hot rolling. Industrially applied rolling parameters were employed for the calculation, and the results can be used as guidance for process optimization in a real production environment.

2. Governing Equations

Thermal boundary loads (heating and cooling) were applied on the rotating roll separated in five main regions, as shown in Figure 1. The roll's surface temperature increases due to heat flux ($\dot{q}_{strip}$) in the contact with the strip, which is hotter compared to the roll during hot rolling. Heat is extracted from the roll due to natural convection and radiation ($\dot{q}_{air}$). Stronger cooling ($\dot{q}_{cool}$) is the effect of the nozzle sprays and the contact with the support roll ($\dot{q}_{WB}$). Between the cooling unit and the hot strip, the coolant flows over the roll, creating different cooling conditions ($\dot{q}_{wet}$). For the definition of the heat flux in the deferent zones, each heat transfer coefficient has to be determined and is described below. The value of the heat transfer coefficient of each thermal load was either found in the literature ($\dot{q}_{air}$), calculated by typical formulations found in the literature ($\dot{q}_{cool}$, $\dot{q}_{WB}$, $\dot{q}_{wet}$) or calibrated using real production data ($\dot{q}_{strip}$).

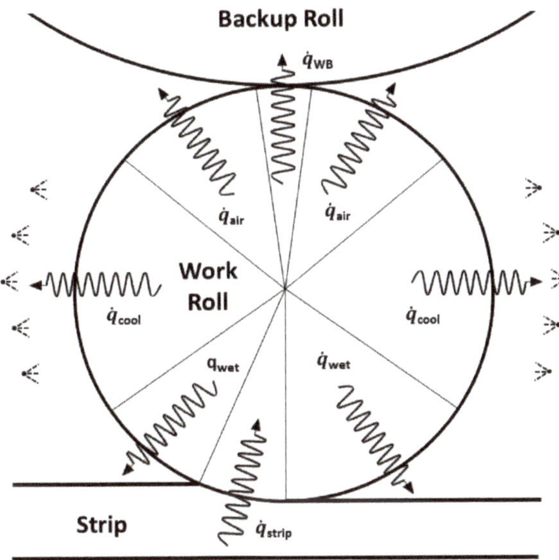

Figure 1. Schematic illustration of the zones with the different boundary conditions.

2.1. Heat Conduction

The basic transient heat conduction equation for cylindrical coordinates and isotropic properties can be written as [12,14,20]:

$$\frac{1}{r}\frac{\partial}{\partial r}\left(k_{wr}\, r\, \frac{\partial T}{\partial r}\right) + \frac{1}{r^2}\frac{\partial}{\partial \varphi}\left(k_{wr}\, r\, \frac{\partial T}{\partial \varphi}\right) + \frac{\partial}{\partial z}\left(k_{wr}\, \frac{\partial T}{\partial z}\right) = \rho\, C_p\, \frac{\partial T_{(t)}}{\partial t} \tag{1}$$

where ρ is the density $(\frac{g}{cm^3})$, C_p the specific heat capacity $(\frac{J}{kg\,K})$, T the transient temperature (K), t the time (s), r is the radial direction (m), φ is the circumferential direction (m), z the longitudinal direction, and k_{wr} the thermal conductivity coefficient $(\frac{W}{m\,K})$. The parameter values are summarized in Table 1. The initial temperature T_{init} was considered homogeneous throughout the roll at the beginning in every case, and can be described by:

$$T_{Roll}(r,\ \varphi,\ m,\ t)\big|_{t=0} = T_{init} \tag{2}$$

where T_{Roll} is the roll temperature.

Table 1. Process parameter constants.

Parameter	Value
$\rho\ [\frac{kg}{m^3}]$	7800
$C_p\ [J\,kgK]$	485
$k_{wr}\ [W\,mK]$	38
$h_{strip}\ [\frac{W}{m^2K}]$	100
$V_{jet}\ [\frac{Liter}{min}]$	3500
$A_{jet}\ [m^2]$	0.4
$P_{jet}\ [Pa]$	0.5×10^6
$T_{coolant}\ [K]$	323
$\mu_{coolant}\ (\frac{Kg}{m\,s})$	4×10^{-4}
$Pr\ [-]$	2
$k_{coolant}\ [W\,mK]$	0.65
$h_{air}\ [\frac{W}{m^2K}]$	10
$T_{air}\ [K]$	300
$\sigma\ [\frac{Kg}{s^3K^4}]$	5.67×10^{-8}
$\varepsilon\ [-]$	0.5

2.2. Boundary Condition for Roll/Strip Contact

The main heat source for the work roll is due to the conduction of heat from the hot strip to the surface of the work roll. The heat flux ($\dot{q}_{Strip}$) $(\frac{W}{m^2})$ depends on the temperature difference between the roll and the strip, and in order to have an effective prediction, a boundary heat transfer coefficient (h_{strip}) $(\frac{W}{m^2K})$ is assumed [12,14,20]:

$$\dot{q}_{strip} = h_{strip}\left(T_{Strip} - T_{Roll}(t)\right) \tag{3}$$

The h_{strip} remained constant through each simulation, and was calibrated and validated in a previous work [1] and implicitly includes the phenomena of frictional heat generation (see Table 1). Similarly, the strip temperature was predicted in a previous work and remained constant through each simulation, with a value equal to the average strip surface temperature between the entry and the exit of the roll bite.

2.3. Boundary Condition for Water Jet Cooling

The most efficient cooling zone on the roll is where water from the jet nozzles comes into contact with the high pressure in the roll. The heat flux boundary condition ($\dot{q}_{cool}$) ($\frac{W}{m^2}$) that applies to this region is as follows [12,14,20]:

$$\dot{q}_{cool} = h_{cool}\left(T_{Roll}(t) - T_{coolant}\right) \tag{4}$$

where h_{cool} is the heat transfer coefficient for the water jet rolling ($\frac{W}{m^2 K}$) and $T_{Roll}(t) - T_{coolant}$ is the transient temperature difference between the roll and the coolant (K). The convective heat transfer coefficient, h_{cool}, can be calculated as follows [20]:

$$h_{cool} = 6870\, W^{0.19}\, P_{jet}^{0.27} \quad (T_{Roll}(t) \le 373\ K) \tag{5}$$

$$
\begin{aligned}
h_{cool} = {}& 29\,10^5\, W^{0.08}\, P_{jet}^{0.05}\, \frac{T_{Roll}(t)-373}{100}\, \frac{B}{T_{Roll}(t)-\, T_{coolant}} \\
& + 6870\, W^{0.19}\, P_{jet}^{0.27}\, \frac{573-T_{Roll}(t)}{100}\quad (T_{Roll}(t) > 373\ K)
\end{aligned}
\tag{6}
$$

where P_{jet} is the nozzle pressure ($\frac{Kg}{m\,s}$), $W = \frac{V_{jet}}{A_{jet}}$ refers to the coolant flow rate per unit area of the coolant on the roll, and $B = \left(\frac{(T_{jet}-373)}{16}\right)^{-0.17}$. The parameter values are summarized in Table 1.

2.4. Boundary Condition for Work Roll/Backup Roll Contact

The boundary condition for the heat flow from the contact between the work roll and the backup roll ($\dot{q}_{WB}$) was calculated by [12,14,20]:

$$\dot{q}_{WB} = h_{WB}\left(T_{Roll}(t) - T_{BRoll}\right) \tag{7}$$

where h_{WB} is the heat transfer coefficient of the contact between the work roll and the backup roll, ($\frac{W}{m^2 K}$), and $T_{Roll}(t) - T_{BRoll}$ is the transient temperature difference between the roll and the backup roll (K).

The heat transfer coefficient between the work roll and backup roll (h_{WB}) can be calculated as:

$$h_{WB} = \frac{1.26\, k_{wr}}{\pi\, D_W\, \sqrt{\pi\, \alpha_{wr}}}\, \sqrt{L_c v_{wr}} \tag{8}$$

where k_{wr} is the work roll thermal conductivity coefficient ($\frac{W}{m\,K}$), D_W is the work roll diameter (m), $\alpha_{wr} = \frac{k_{wr}}{\rho\, C_p}$ is the work roll thermal diffusivity ($\frac{m^2}{s}$), L_c is the contact length between the rolls (m) and v_{wr} is the work roll rotational velocity ($\frac{m}{s}$). The contact length, L_c, was considered equal to the length of the backup roll. The thermal characteristics are summarized in Table 1 and the geometrical in Table 2.

Table 2. Work roll geometry and physical properties.

Parameter	Value
Work Roll Length (m)	3
Work Roll Diameter (m)	0.95
Support Roll Length (m)	2.5
Poisson's ratio	0.3
Young's modulus [GPa]	200
Thermal expansion coefficient [$\frac{1}{K}$]	12.5×10^{-6}

2.5. Boundary Condition for Wet Surface between the Cooling Units and Roll Bite

The roll surface below the cooling units is in contact with coolant flowing over the roll surface under atmospheric pressure. The boundary condition that describes that region is [12,14,20]:

$$\dot{q}_{wet} = h_{wet} \left(T_{Roll}(t) - T_{coolant} \right) \tag{9}$$

where h_{wet} is the heat transfer coefficient for the wet surface below the cooling unit $\left(\frac{W}{m^2 K} \right)$ and $T_{Roll}(t) - T_{coolant}$ is the transient temperature difference between the roll and the coolant (K). The h_{wet} can be calculated by [20]:

$$h_{wet} = 0.023 \left(\frac{v_w \, l_c}{\mu_{coolant}} \right)^{0.8} Pr^{0.4} \frac{k_{coolant}}{l_c} \tag{10}$$

where v_{wr} is the work roll rotational velocity $\left(\frac{m}{s} \right)$, l_c is the length of the coolant contact area (m), $\mu_{coolant}$ is the viscosity of the coolant $\left(\frac{Kg}{m\,s} \right)$, Pr is the Prandl number (–) and $k_{coolant}$ is the thermal conductivity coefficient of the coolant $\left(\frac{W}{m\,K} \right)$. The length of the coolant contact area was considered to be equal to the work roll length. The parameter values are summarized in Table 1.

2.6. Boundary Condition for Air Cooling

In the region between the cooling units and the work roll, to support roll contact, heat is lost from the surface of the work roll, assuming convection to air and radiation [12,14,20]:

$$\dot{q}_{air} = h_{air} \left(T_{Roll}(t) - T_{air} \right) + \sigma \, \varepsilon \, A \left(T_{Roll}(t)^4 - T_{air}^4 \right) \tag{11}$$

where h_{air} is the heat transfer coefficient of the air $\left(\frac{W}{m^2 K} \right)$, $T_{Roll}(t) - T_{air}$ the transient temperature difference between the roll and the air (K), σ is the Stephan–Boltzmann radiation constant $\left(\frac{kg}{s^3 K^4} \right)$, A is the area of the emitting body (m^2) and ε is the emissivity of steel. The parameter values are summarized in Table 1.

3. Model Description

A three-dimensional finite element model was designed to study the temperature and thermal camber evolution during the hot rolling process of a 5754 aluminum alloy strip under industrial service conditions, taking into consideration also the heat flux in the longitudinal direction. Thermal camber is the difference in radius between the center of the roll (R_{center}) and the position in width where the edge is located (R_{edge}), excluding the last 20 mm to avoid edge effects. Camber can be described as:

$$Camber = R_{center} - R_{edge} \tag{12}$$

The direct thermo-mechanical coupling allowed the thermal balance to be achieved in each timestep, and subsequently the thermal expansion to be calculated and each node position updated, resulting in new roll geometry. The simplified form of the description of the thermal expansion equation can be represented in the following form:

$$\frac{\Delta L}{L} = \alpha \, \Delta T \tag{13}$$

where α is the coefficient of thermal expansion $\left(\frac{m}{m\,K} \right)$, L is length (m) and T is the temperature (K).

For this purpose, a work roll was modeled on LS-DYNA software with a Lagrangian formulation and 3D mesh with an element size of 10 mm close to the surface of the roll where the largest temperature gradients develop. Due to the rolls' symmetry, the examination of only half of the roll reduced the computational effort without limiting the result quality. The roll was considered elastically deformable, and the geometry and physical properties of the roll [1,14] are shown in Table 2.

The thermal loads described in Section 2 remained constant through each pass. In this study, the effects of strip temperature, roll temperature, rolling speed and cooling unit activation strategy were taken into consideration.

4. Results

A representative snapshot of the temperature gradient on the work roll from a characteristic experiment is illustrated in Figure 2. The red color depicts the highest temperature region on the roll surface while in contact with the high temperature strip. Soon after the roll surface disengages from the strip, it quickly cools down, passing through the cooling units, reaching the same temperature as when in the steady state before the contact with the strip. The width of the thermally affected region is similar to the width of the strip.

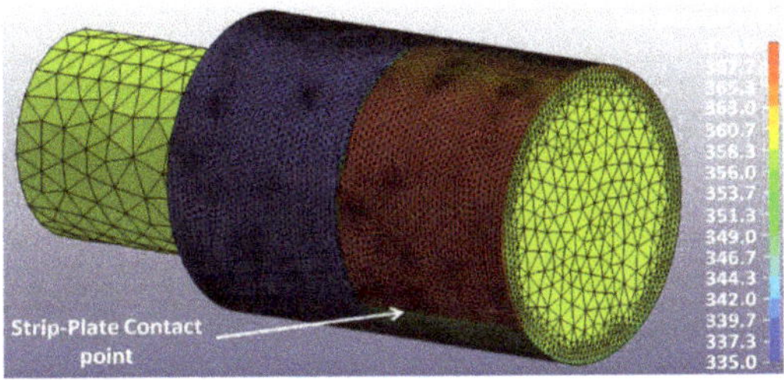

Figure 2. Roll temperature gradient during rolling.

4.1. Validation

The model can only be considered reliable after a validation process takes place. Validation can be very difficult in the aggressive industrial environment that also has various limitations. Such an example is the dynamic temperature measurement on the roll surface due to high temperatures, limited space and water/oil mist that limits the accuracy of the measurements. Additionally, the dynamic measurement of the roll diameter during the pass schedule is out of question. Even the static measurement of the roll can be very difficult because the roll has to be removed from the mill, which requires enough time to finally make the measurement unrealistic. The validation was based on the assumption that the variation in the roll geometry is reflected on the final strip geometry evolution, something that can easily and reliably be measured by the rolling mills gauge measurement system. Thus, the difference between the strip crown at the beginning and at the end of a predesigned industrial pass trial was compared with the roll camber difference between the beginning and the end of the same pass. An indicative comparison between the measured crown evolution (Figure 3) and the calculated camber evolution of the same pass (Figure 4) is presented, which has assisted in the validation of the model. The same validation sequence was conducted in several different passes, which were in good agreement regarding accuracy with the presented results, before the model was considered reliable. The result of the original data recording from the online thickness measuring device can be seen in Figure 3. The data refer to a pass with reduction from 6.6 mm to 4.4 mm, with entry temperature of 600 K and exit temperature of 583 K. The average separating force for the pass was 5.8 MN and the speed was 60 rpm. The roll temperature was measured before the pass to be 340 K.

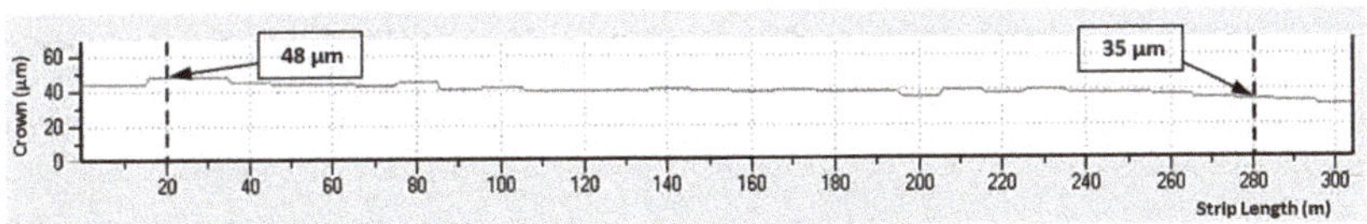

Figure 3. Original record of crown from the online thickness measuring device.

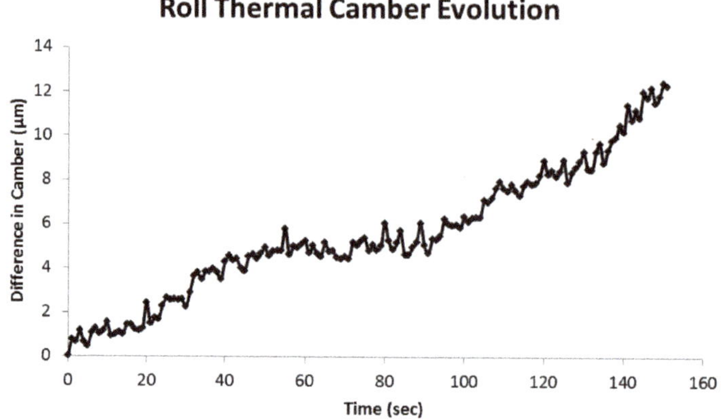

Figure 4. Calculated roll thermal camber evolution for the validation pass.

The original data were compared with the simulation result of the same pass. The crown evolution simulation result is illustrated in Figure 4. The first and last 20 m were omitted from each case to avoid any instability and local effects; however, the thermal load on the first 20 m was included in the calculation.

As can be seen in Figure 3, the crown difference between the two reference points was 13 μm, which is in good agreement with the simulated camber evolution, which was calculated to be 12.3 μm for the same pass (Figure 4).

4.2. Analysis of Temperature

For the same validation pass, the temperature evolution is analyzed in Figure 5. It can be observed that although the temperature on the surface (red line) comes into a steady state after approximately 80 s, 15–60 mm (remaining lines) below the surface the temperature continues to rise even at the end of the pass. The roll does not reach thermal balance even at a long pass, which justifies the geometric variation through process.

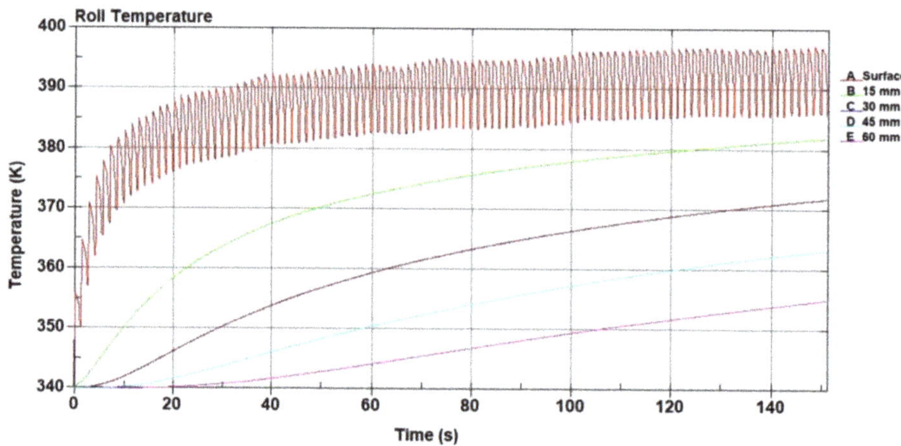

Figure 5. Temperature evolution across the roll surface.

4.3. Comparison of Cooling Unit Efficiency

During the last passes, cooling only at the entry side is applied on the roll, as can be seen in Figure 6. In the above-mentioned validation step, only the entry cooling was applied as well. However, the mill can utilize cooling also to the exit side, which in most cases is not used due to limitations related to surface quality and stains. Of course, the application of cooling from both sides is more efficient compared to the entry side only, and the benefit will be ~30% according to the simulation results of interfacial heat transfer presented in Figure 7.

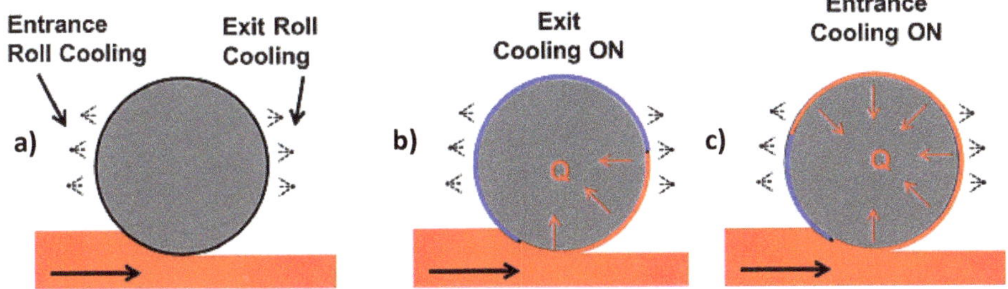

Figure 6. Schematic representation of the position of the cooling units (**a**) and the influence of only exit cooling (**b**) and only entry cooling (**c**) on the energy balance.

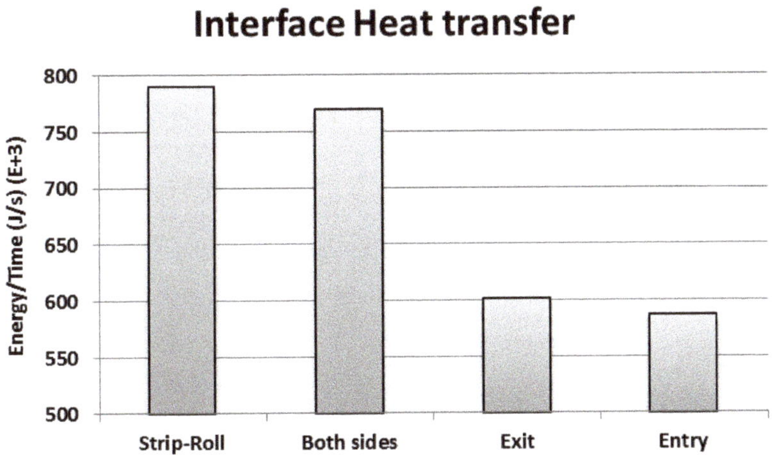

Interface Heat transfer

Figure 7. Comparison of heat flux between the roll and the strip for different cooling strategies.

This cooling efficiency difference is imprinted on the geometric evolution of the roll through the pass. In Figure 8, the FEM results can be seen, which compare the camber evolution between entry cooling applied and entry and exit cooling applied simultaneously. After 100 s of processing, the camber increased by 6 μm in the case of only entry cooling, and by 4 μm when cooling from both sides was applied.

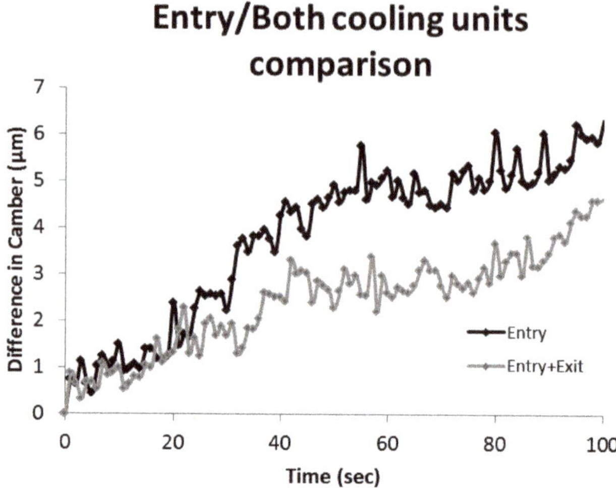

Figure 8. Comparison of camber evolution between entry cooling applied and entry and exit cooling applied simultaneously.

4.4. Influence of Initial Roll Temperature

Very important as well is the influence of the roll's initial temperature on the camber evolution, as can be seen in Figure 9. After 100 s of processing the camber increased by ~6 μm in the case of a 340 K roll initial temperature, and by ~4 μm in the case of the 380 K roll initial temperature.

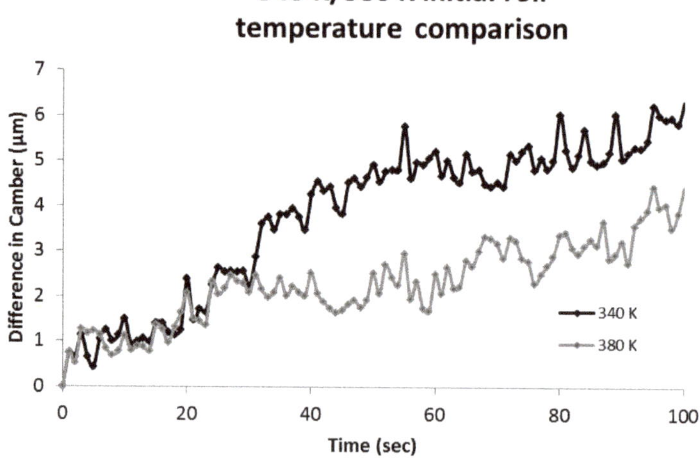

Figure 9. Comparison of camber evolution between rolls with 340 K και 380 K initial temperature.

The rolls at higher temperatures, apart from being geometrically more stable, were also thermally more stable. Figure 10 shows the temperature evolution diagram for the 340 K initial roll temperature and for the 380 K initial roll temperature. For 100 s of processing and at 15 mm away from the surface, the temperature difference between the start and the finish of the pass according to the simulation experiment was 38 K ($\Delta T340K = 340 - 378 = 38$ K) for the cold roll and 13 K ($\Delta T380K = 380 - 393 = 13$ K) for the hot one. Very interesting is the fact that although the temperature difference at the beginning of the simulation experiment between the rolls was 40 K, the temperature after 100 s was very close, with only 5 K difference in surface temperature.

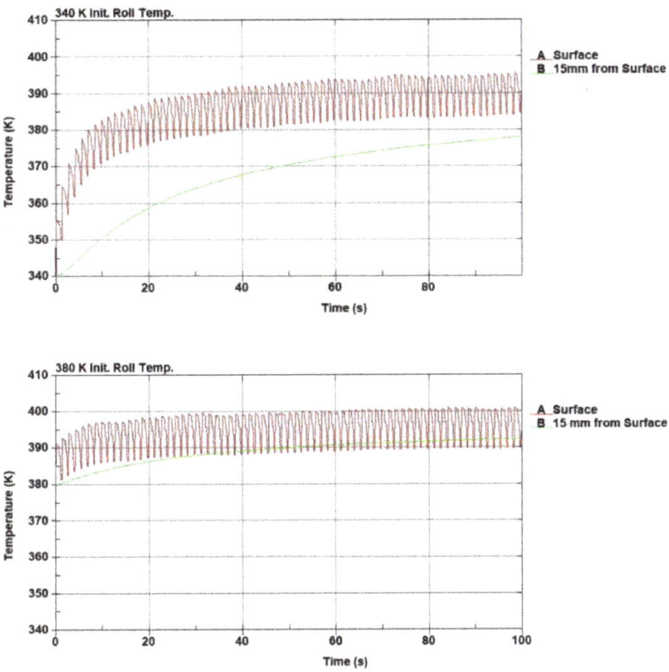

Figure 10. Comparison of temperature evolution between rolls with 340 K (**up**) and 380 K (**down**) initial temperature.

The hot roll had slightly lower heat dissipation from the strip to the roll (~1%) during the steady state compared to the cold roll, however it had significantly lower heat dissipation at the beginning of the pass (~7%) (Figure 11). The lower heat dissipation should balance the temperature difference between the front and back end of the strip compared to the main body. This will further improve the flatness of the strip through the length as the temperature variation results in rolling force variation, and thus crown variation.

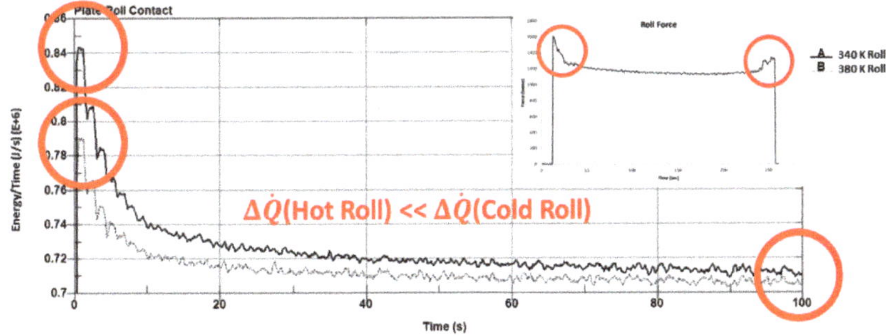

Figure 11. Comparison of heat dissipation from the strip towards the roll for different initial roll temperatures (340 K and 380 K).

4.5. Influence of Strip Temperature

Although the strip temperature is also a factor that might be able to alter the roll camber, as can be seen in Figure 12, the influence is not as significant compared to the above-mentioned process parameters. A higher strip temperature will further increase the surface temperature, but at the same time it will increase the temperature difference between the roll surface and the coolant, which will increase the cooling units' efficiency.

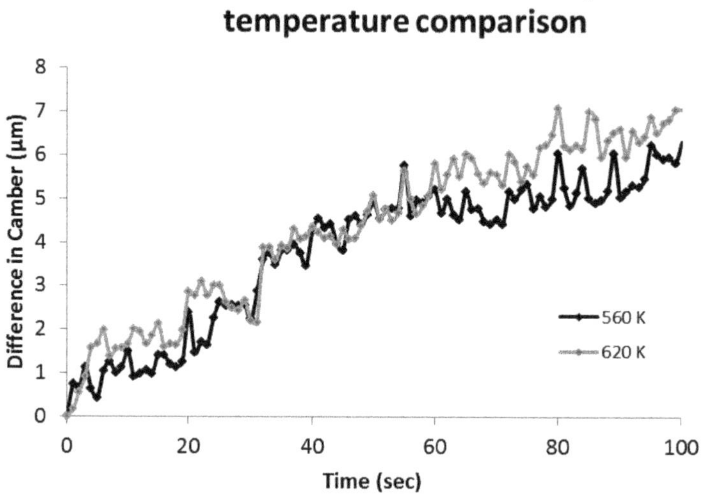

Figure 12. Comparison of roll camber evolution for rolling strips with different temperatures (560 K and 620 K).

5. Discussion

In the final passes of the pass schedule, higher forces can be observed at the front and the end of the strip because of the lower temperature reaching a steady state in between. Moreover, the roll camber evolves due to thermal loads. Consequently, the resulting crown of the strip is significantly higher at the front and back compared to the main body, where the crown has, in many cases, a decreasing trend through length. The variation as well as the absolute value of the crown can sometimes be so high that the whole strip or parts of it will be out of specification, or it will make difficult any further processing.

Some modern rolling mills are equipped with automatic roll bending actuators to compensate for any variation in crown due to force inconsistency and thermal camber through the length of the strip, but often the control window is very narrow so as to correct a poor pass schedule design. Moreover, the thinner the strip, the more difficult it is for the material to flow laterally, and corrections can result in waviness on certain parts of the strip. The same problem will be even stronger in cases where a strip with inconsistent crown reaches the cold rolling stage. For this reason, the proper control of the thermal state and the roll camber with the optimum cooling strategy is crucial for the flatness of the final product.

The effects of the phenomena taking place during rolling and of the process parameters on the thermal camber and the temperature evolution of the roll are difficult to understand and quantify. The proposed model was designed to enhance the understanding of the process and assist in assessing the corrective measures for improvements. Taking into consideration that in the industrial environment the most common methodology for process improvement is "trial and error", the proposed model will enable us to investigate different parameters unconditionally with substantially lower cost in comparison to industrial trials. The simulation results provide important knowledge gain and guidance for the determination of the proper rolling conditions, which can increase the success rate of the following industrial trials with a minimal rolling trial optimization sequence.

The thermal analysis proved that the roll does not reach a total thermal steady state, a condition where the camber remains constant, and also pointed out the temperature whereat the steady state is reach on the surface. The approach for improvement was to reduce the steady state temperature by increasing the cooling efficiency, to bring the roll closer to this thermal steady state by increasing the roll temperature, and to quantify the improvements for prioritization purposes. The influence of the strip temperature was studied as well.

The model showed the improvement of the application of cooling from both sides of the work roll. The cooling efficiency was improved by ~30%, with similar improvement in the thermal camber evolution. However, the challenges of the use of cooling at the exit side of the roll are well known. Most of the time, these are related to the inefficient removal of the emulsion falling on the strip from the cooling unit at the exit side. Especially in the last passes of a reversing mill, where the strip is coiled, it can cause stains that will diminish the strip's final surface quality. Of course, with the suitable equipment and a very careful control of the process, such problems can be avoided. Although it can be avoided, the additional effort and the idea of the process being susceptible to visual defects on the final product often leads to the use of the cooling unit only from the entrance side. For products in which the appearance is the basic concern, such an approach is enough, but in higher value-adding products in which flatness is also very significant, the geometrical control of the work roll becomes very important.

On the other side, the increase in the average roll temperature will result in a more geometrically stable roll. This is due to the fact that the temperature difference between both the roll and the strip, but also between the roll surface and roll center, will become smaller, deteriorating the heat flux. According to the simulation, an increase of 40 K in average roll temperature will result in the significant improvement of thermal camber evolution by reducing the finishing camber by 2 μm. Effort is required to increase the rolls' core temperature. The concern pertains to the effect of a hotter roll on the steady state temperature of the roll surface during rolling. From simulation findings, it appears that the average temperature of the roll has almost no impact on the steady state temperature at the surface of the roll.

In an industrial reversing mill pass schedule, where the time for which the strip is in contact with the roll can even be less than half, the way to achieve such a temperature increase is, firstly, to continue rolling for several pass schedules, but, more importantly, to avoid roll cooling during the inter-pass time. Even after several pass schedules, if the roll is cooled during the inter-pass time, it will remain relatively cold, and as a result, will never reach a steady state and will be geometrically unstable on high thermal loads and long passes.

An additional benefit of a hot roll is seen during its first contact with the strip. A common problem in rolling is that the first and last meters of the strip have lower temperatures compared to the main body of the strip, which results in an increase in separation forces and, consequently, in poor flatness and properties variation. The lower temperature can be attributed to many reasons, such as the contact of the strip with the cold core in the center during coiling, the free surfaces at the front and back end, the lower speed with which each pass starts before reaching the maximum speed, etc. The smaller temperature difference between a higher temperature roll and the strip will cause a drop of heat flux between the strip and the roll, and more heat will be maintained in the front and the back of the strip. In some cases, in order to further enhance this effect, a practice is to start the pass with turned off cooling units in order for the roll surface to more quickly reach the steady state temperature, and then the cooling is applied at a few meters from the beginning of the pass.

The strip sticking to the roll is a common problem in aluminum hot rolling. Apart from other issues, it can be related to either very high or very low roll surface temperatures. The surface temperature can mainly be controlled by the effective cooling. In the cases wherein sticking is observed, the application of cooling at the exit side becomes again very important. To cool the roll between the passes in order to start a certain pass with the lowest temperature possible, or to avoid a high temperature roll core due to the fear that it will impact the final product's surface quality, are not recommended.

6. Conclusions

The strip profile is significantly influenced by the temperature profile on the roll and the camber evolution caused by thermal expansion. The model was validated by comparing industrial measurements of crown evolution through length with the calculated evolution of the roll camber. The conclusions are summarized below.

The application of cooling from both the entry and the exit side is ~30% more efficient compared to entry side cooling only. The final camber evolution at the end of a 100 s pass will drop from 6 μm in the case of only entry cooling to 4 μm for both entry and exit cooling.

The initial roll temperature will affect the final camber substantially. The hotter roll will be geometrically more stable compared to the colder roll. More specifically, for a 100 s pass the camber increases by ~6 μm in the case of 340 K roll initial temperature, and by ~4 μm in the case of the 380 K roll initial temperature.

In a hotter roll the surface temperature will reach a steady state faster compared to a colder one, shaping a different temperature gradient in the strip as well.

The surface temperature of the roll during the process has only slight dependency on the roll's average temperature. A high increase in roll temperature will lead to only a very small increase in surface temperature.

The strip temperature will alter the thermal camber, however it has a weaker effect compared to the average roll temperature and cooling strategy.

Author Contributions: Conceptualization, E.G.; methodology, E.G. and S.P.; software, E.G.; validation, E.G.; formal analysis, E.G. and S.P.; investigation, E.G.; resources, S.P.; data curation, E.G.; writing—original draft preparation, E.G.; writing—review and editing, S.P.; visualization, E.G.; supervision, S.P.; project administration, S.P. All authors have read and agreed to the published version of the manuscript.

Funding: This research received no external funding.

Acknowledgments: The authors acknowledge the technical management of ELVAL S.A. for providing access to the production site and for the permission to utilize production data for the validation of our model. The discussions with A. Mavroudis, M. Gonidakis and D. Kortselis throughout the execution of the project are also highly appreciated. The authors also express gratitude to the Hellenic Research Centre for Metals—ELKEME S.A. for their support in this research.

Conflicts of Interest: The authors declare no conflict of interest.

References

1. Gavalas, E.; Papaefthymiou, S. Prediction of Plate Crown during Aluminum Hot Flat Rolling by Finite Element Modeling. *J. Manuf. Mater. Process.* **2019**, *3*, 95. [CrossRef]
2. Guo, R.M. Determination of Optimal Work Roll Crown for a Hot Strip Mill. *Iron Steel Technol.* **1989**, *66*, 52–60.
3. Sikdar, S.; Shylu, J.; Pandit, A.; Dasu, R. Analysis of roll stack deflection in a hot strip mill. *J. Braz. Soc. Mech. Sci. Eng.* **2007**, *29*. [CrossRef]
4. Steinboeck, A.; Ettl, A.; Kugi, A. Dynamical Models of the Camber and the Lateral Position in Flat Rolling. *Appl. Mech. Rev.* **2017**, *69*, 040801. [CrossRef]
5. Fukushima, S.; Washikita, Y.; Sasaki, T.; Nakagawa, S.; Buei, Y.; Yakita, Y.; Yanagimoto, J. *High-Accuracy Profile Prediction Model for Mixed Scheduled Rolling of High Tensile Strength and Mild Steel in Hot Strip Finishing Mill*; Nippon Steel & Sumitomo Metal Technical Report; Nippon Steel Corporation: Tokyo, Japan, March 2016.
6. Wang, T.; Xiao, H.; Zhao, T.Y.; Qi, X.D. Improvement of 3-D FEM Coupled Model on Strip Crown in Hot Rolling. *J. Iron Steel Res. Int.* **2012**, *19*, 14–19. [CrossRef]
7. Shigaki, Y.; Montmitonnet, P.; Silva, J.M. 3D finite element model for roll stack deformation coupled with a Multi-Slab model for strip deformation for flat rolling simulation. *AIP Conf. Proc.* **2017**, *1896*, 190018. [CrossRef]

8. Atack, P.A.; Robinson, I.S. An investigation into the control of thermal camber by spray cooling when hot rolling aluminium. *J. Mater. Process. Technol.* **1994**, *45*, 125–130. [CrossRef]

9. Tseng, A.A.; Tong, S.X.; Chen, T.C. Thermal Expansion and Crown Evaluations in Rolling Processes. *Mater. Des.* **1997**, *17*, 193–204. [CrossRef]

10. Stürmer, M.; Dagner, J.; Manstetten, P.; Köstler, H. Real-time simulation of temperature in hot rolling rolls. *J. Comput. Sci.* **2014**, *5*, 732–742. [CrossRef]

11. Jiang, M.; Li, X.; Wu, J.; Wang, G. A precision on-line model for the prediction of thermal crown in hot rolling processes. *Int. J. Heat Mass Transf.* **2014**, *78*, 967–973. [CrossRef]

12. Ginzburg, V.B.; Bakhtar, F.A.; Issa, R.J. Application of Coolflex model for analysis of work roll thermal conditions in hot strip mills. *Iron Steel Eng.* **1997**, *74*, 38–45.

13. Lin, Z.-C.; Chen, C.-C. Three-dimensional heat-transfer and thermal-expansion analysis of the work roll during rolling. *J. Mater. Process. Technol.* **1995**, *49*, 125–147. [CrossRef]

14. Abbaspour, M.; Saboonchi, A. Work roll thermal expansion control in hot strip mill. *Appl. Math. Model.* **2008**, *32*, 2652–2669. [CrossRef]

15. Guo, Z.F.; Li, C.S.; Xu, J.Z.; Liu, X.H.; Wang, G.D. Analysis of temperature field and thermal crown of roll during hot rolling by simplified FEM. *J. Iron Steel Res. Int.* **2006**, *13*, 27–30. [CrossRef]

16. Benasciutti, D.; Brusa, E.; Bazzaro, G. Finite element prediction of thermal stresses in work roll of hot rolling mills. *Procedia Eng.* **2010**, *2*, 707–716. [CrossRef]

17. Li, C.-S.; Yu, H.-L.; Deng, G.-Y.; Liu, X.-H.; Wang, G.-D. Numerical Simulation of Temperature Field and Thermal Stress Field of Work Roll during Hot Strip Rolling. *J. Iron Steel Res. Int.* **2007**, *14*, 18–21. [CrossRef]

18. Trull, M.; McDonald, D.; Richardson, A.; Farrugia, D. Advanced finite element modelling of plate rolling operations. *J. Mater. Process. Technol.* **2006**, *177*, 513–516. [CrossRef]

19. Bao, L.; Qi, X.-W.; Mei, R.-B.; Zhang, X.; Li, G.-L. Investigation and modelling of work roll temperature in induction heating by finite element method. *J. Sout. Afr. Inst. Min. Metall.* **2018**, *118*, 735–743. [CrossRef]

20. Deng, G.Y.; Zhu, H.T.; Tieu, A.K.; Su, L.H.; Reid, M.; Zhang, L.; Wei, P.T.; Zhao, X.; Wang, H.; Zhang, J.; et al. Theoretical and experimental investigation of thermal and oxidation behaviours of a high speed steel work roll during hot rolling. *Int. J. Mech. Sci.* **2017**, *131–132*, 811–826. [CrossRef]

Publisher's Note: MDPI stays neutral with regard to jurisdictional claims in published maps and institutional affiliations.

Article

Highly Enhanced Hot Ductility Performance of Advanced SA508-4N RPV Steel by Trace Impurity Phosphorus and Rare Earth Cerium

Yu Guo [1], Yu Zhao [2] and Shenhua Song [1,*]

[1]   Shenzhen Key Laboratory of Advanced Materials, School of Materials Science and Engineering, Harbin Institute of Technology, Shenzhen 518055, China; guoyu@stu.hit.edu.cn

[2]   Anhui Key Laboratory of High-Performance Non-Ferrous Metal Materials, Anhui Polytechnic University, Wuhu 241000, China; zhaoyu@ahpu.edu.cn

*   Correspondence: shsong@hit.edu.cn; Tel.: +86-755-2603-3465

Received: 20 September 2020; Accepted: 24 November 2020; Published: 28 November 2020

Abstract: Advanced SA508-4N RPV steel samples, unadded, P-added, and P+Ce-added, are investigated on their hot ductility behavior. Hot tensile tests are carried out in the temperature range of 750 to 1000 °C through a Gleeble 1500D machine. It is demonstrated that the deformation temperatures of all the three steels are located in the austenite single-phase region. There is no ductility trough present for the P+Ce-added steel, but the unadded one exhibits a deep ductility trough. The reduction of area (RA) of the former is always higher than 75% and increases with rising temperature until reaching ~95% at 900 °C or above, whereas the lowest RA value of the latter is only ~50% at 850 °C. Microanalysis indicates that the grain boundary segregation of P and Ce takes place in the tested P+Ce-added steel. This may restrain the boundary sliding so as to improve the hot ductility behavior of the steel. Furthermore, the addition of P and Ce is able to facilitate the occurrence of the dynamic recrystallization (DR) of the steel, lowering the initial temperature of DR from ~900 to ~850 °C and thereby enhancing the hot ductility performance. Consequently, the combined addition of P and Ce can significantly improve the hot ductility of SA508-4N RPV steel, thereby improving its continuous casting performance and hot workability.

Keywords: hot ductility; reactor pressure vessel steel; grain boundary segregation; dynamic recrystallization; grain boundary sliding

---

1. Introduction

The reactor pressure vessel (RPV) is a critical component in nuclear power plants, relating to the whole life of the plant due to its irreplaceability [1,2]. Due to its severe service condition, some superior performances of the RPV steel are attracting more and more attention [2,3]. SA508-3 steel mainly alloyed with Mn and Mo has been widely used in the nuclear industry for more than 30 years. However, this steel is unable to satisfy future applications due to its insufficient hardenability and irradiation embrittlement [4]. Nowadays, plenty of research studies focus on a new generation of RPV steel (i.e., SA508-4N steel mainly alloyed with Ni, Cr, and Mo). According to ASTM (American Society for Testing and Materials), advanced SA508-4N steel has higher strength, better fracture toughness, and outstanding hardenability as compared with SA508-3 steel [5–7]. As reported by Lee et al. [5], the Charpy impact toughness and fracture toughness of SA508-4N steel were evidently improved because of the addition of Ni and Cr. They also found that the yield strength and upper shelf energy of the steel were enhanced by increasing the tempered martensite phase fraction controlled by the cooling rate [6].

The process of continuous casting and rolling, which has the advantage of uncomplicatedness and cost-effectiveness, is widely employed in the manufacture of many steel products [8]. However, if the hot ductility (HD) of a steel is not high enough (less than 60% reduction of area), transverse cracking could take place during continuous casting, causing severe damage in the quality of products [9]. The poor HD of a steel also affects its hot working performance, such as hot rolling or forging. Moreover, the accumulation of some residual impurity elements during the recycling of steel scraps would severely deteriorate the HD of steel [9,10]. As demonstrated in [11], SA508-4N steel exhibits a relatively poor HD performance between 750 and 900 °C, especially at around 850 °C, where the reduction of area (RA) is much lower than 60%. As a result, for this RPV steel, the straightening operation of continuous casting must be conducted above 900 °C, and the hot working must also be carried out above this temperature. No doubt, it would be difficult to conduct these operations in fact as the temperature window is narrow. In order to improve the quality and productivity of continuous casting and hot working of this steel, it is necessary to raise its hot ductility, especially between 800 and 900 °C.

Recent studies demonstrated [9,11–16] that phosphorus segregated at grain boundaries (GBs) had a positive influence on the HD of some steels through restraining carbide precipitation along the boundaries. Usually, the carbide precipitation is very likely to form at high temperatures in steels containing some strong carbide-forming alloying elements, such as V, Ti, and Nb. Mintz et al. [12] reported that P could enhance the HD of Nb-bearing steels by suppressing the precipitation of NbC and thus restraining the GB sliding. Besides, a study by Jiang et al. [13] indicated that the HD of 1Cr0.5Mo steel could be significantly enhanced by adding 0.054 wt.% P. Further study suggested that P tended to strengthen the thin ferrite layers formed at austenite grain boundaries so that the difference in strength between the austenite matrix and ferrite layer could be reduced, and consequently, the hot deformation could occur more uniformly, enhancing the HD accordingly. Recently, Guo et al. [11] found that P could effectively enhance the HD of SA508-4N RPV steel through grain boundary segregation (GBS). It is proposed that the P atoms segregated at the GBs might suppress the GB sliding by occupying the vacant sites at the boundaries. As a result, the RA of the P-doped steel is considerably raised, making the HD trough much shallower. However, it is acknowledged that P could also seriously deteriorate some other properties of the steel, especially toughness [17–19]. The GBS of P could raise the ductile-to-brittle transition temperature (DBTT) by weakening the boundary, which significantly affects the applicability of steel. Therefore, it is not recommended to improve the HD of a steel by the addition of P.

In many cases, rare earths (REs) have beneficial influences on the mechanical properties of steels, such as toughness and hot ductility [20–26]. The enhancement mechanisms of toughness and hot ductility are mainly grain refinement and GBS of REs. Liu et al. [21] found that Ce could considerably improve the impact toughness of an industrial low-carbon steel in three ways: reducing pearlite, refining microstructure, and cleaning GBs. Meanwhile, the GBS of Ce is conducive to lowering the DBTT of the welding heat-affected zone in SA508-3 RPV steel [22]. A study by Jiang and Song [23] indicated that Ce played an important role in the HD performance of 1Cr-0.5Mo steel. The trace addition of Ce caused a remarkable rise in the RA of the steel due to the GBS of Ce by restraining the GB sliding and facilitating the dynamic recrystallization (DR). Ma et al. [24] found that RE (Ce or La) could enhance the hot workability behavior of duplex stainless steel mainly by reducing the difference of hardness between ferrite and austenite and promoting DR. Guo et al. [25] also demonstrated that the HD of SA508-3 RPV steel could be significantly improved by the GBS of Ce. In addition, the theoretical calculations [27,28] also showed that the addition of RE elements in the steel could increase the intergranular cohesion, thus strengthening the GB.

According to the above-mentioned information, it is envisaged that the combined addition of P and Ce in the SA508-4N RPV steel could improve the HD behavior of the steel, and at the same time, the P-induced low-temperature embrittlement might be suppressed. In this study, a Gleeble machine together with various characterization techniques was employed to investigate in detail the combined impact of P and Ce on the HD behavior of SA508-4N RPV steel, demonstrating that the HD performance of the RPV steel could be highly ameliorated, thus eliminating the HD trough exhibited by the unadded steel. In order to better explore the combined effect of P and Ce, the results for the

P-added SA508-4N steel reported in our previous work [11] were also incorporated into the present study for comparison.

2. Materials and Methods

Three heats of experimental SA508-4N RPV steel were melted by vacuum induction, cast into a sand mold, and finally hot-forged into plates after homogenization annealing (China Iron and Steel Research Institute Group, Beijing, China). The chemical composition of the steels is listed in Table 1. Clearly, they are unadded, added with 0.043 wt.% P, and added with 0.052 wt.% P and 0.030 wt.% Ce. In short, the three steels are hereafter called "unadded steel", "P-added steel", and "P + Ce-added steel".

Table 1. Chemical Composition of Experimental Steels (wt.%).

Component	C	Si	Mn	P	S	Ni	Cr	Mo	Ce	Fe
Unadded	0.14	0.16	0.25	0.0022	0.0033	2.98	1.45	0.51	-	Bal.
P-added	0.14	0.18	0.25	0.043	0.0032	3.08	1.80	0.50	-	Bal.
P+Ce-added	0.14	0.17	0.25	0.052	0.0033	3.02	1.80	0.50	0.030	Bal.

A Gleeble 1500D machine (Data Sciences International, Saint Paul, MN, USA) was employed to carry out the hot tensile tests. The tests were operated under a vacuum of 20–30 Pa. The size of specimens was cylindrical with 10 mm diameter and 120 mm gage length (uniform heating region). The specimens were heated up to 1300 °C and held there for 180 s, followed by cooling down to different temperatures (750–1000 °C) at a cooling speed of $5\,^{\circ}\mathrm{C}\,\mathrm{s}^{-1}$ [13]. After holding for 10 s at each temperature, the specimens were deformed until fracture with a rate of $10^{-3}\,\mathrm{s}^{-1}$ [29], cooled to 300 °C under the same vacuum condition, and subsequently air-cooled to room temperature. In order to estimate the possibility of transverse cracking during continuous casting, the RA is usually used [30]. If the RA is less than 60%, the transverse cracking could occur during the straightening operation of continuous casting [31]. Hence, the RA was measured to evaluate the HD of the steel. Besides, the fracture surfaces were examined using scanning electron microscopy (SEM, Hitachi S-4700, Hitachi, Tokyo, Japan). Optical metallography (OM, HOMA2000, XPK, Shenzhen, China) and electron backscatter diffraction (EBSD, Hikarui, EDAX, Mahwah, NJ, USA) were used to characterize microstructural features.

Field-emission scanning transmission electron microscopy (FE-STEM, JEM-2100F, JEOL, Tokyo, Japan) coupled with energy dispersive X-ray spectroscopy (EDS, Oxford INCA, Oxford, UK) was adopted to further investigate microstructural characteristics and microchemistry at the GBs [11]. The field emission gun in the FE-STEM system may provide a large beam current in a probe with a size of only 0.5–1 nm [32]. The disc samples for transmission electron microscopy were 3 mm in diameter, prepared by mechanically polishing and dual-jet electropolishing. For microanalysis, five GBs were chosen, and more than five positions on each GB were analyzed. The mean value of data points gained was used as the observed result. Further description of GB microanalysis by FE-STEM can be seen elsewhere [33].

3. Results

3.1. Hot Ductility

The HD curves of the steels are shown in Figure 1. It is clear that there are ductility troughs in the temperature range of 750–900 °C for both the unadded and P-added steels, but the trough is apparently shallower for the P-added steel. As a whole, the RA values of the P-added steel are all more than 70%, which are always higher than those of the unadded steel. The RA of the P-added steel is about 75% at 750 °C and then goes down to about 70% at 800 and 850 °C. Nevertheless, the RA value of the unadded steel deformed at 750–850 °C is 70–50%, attaining the lowest RA of ~50% at 850 °C. From 850 to 900 °C, the RA value for both the unadded and P-added steels increases rapidly to higher than 90%. On the other hand, there is no ductility trough present for the P+Ce-added steel so that it exhibits an outstanding HD behavior at all deforming temperatures. It is obvious that the RA of the P+Ce-added steel increases continuously with increasing temperature. Starting from ~75% at 750 °C, it rises up to ~90% at 850 °C, where it is just approximately 70% and 50%, respectively, for the P-added and unadded steels. Above 900 °C, the HD has no obvious difference between these steels, whose RA values are all higher than 90%. Therefore, the combined addition of P and Ce in the SA508Gr.4N RPV steel is able to achieve a superior HD performance.

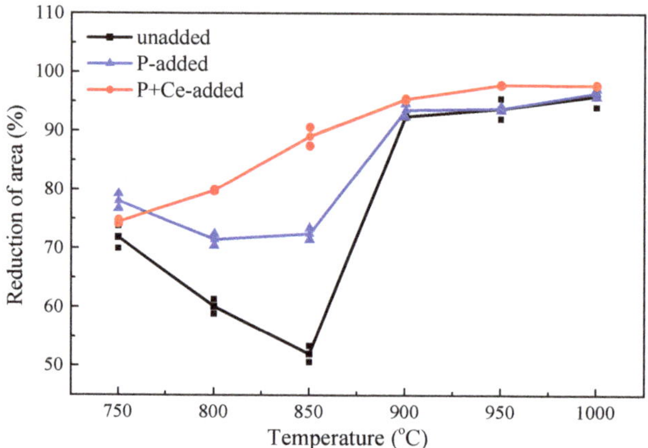

Figure 1. Hot ductility curves for the unadded, P-added, and P+Ce-added steels.

3.2. Fractographic Analysis

SEM micrographs of the fracture surfaces for the unadded, P-added, and P+Ce-added samples are represented in Figure 2. When deformed at 750 °C, three types of samples all exhibit similar ductile fracture with a number of small dimples (Figure 2a–c). This is well in agreement with their similar high RA values (Figure 1). When deformed at 800 °C, the fracture morphology for the unadded sample is ductile with some brittle characteristics demonstrated seemingly by some intergranular facets (Figure 2d). During the tensile test, some cracks could initiate and propagate along austenite grain boundaries. This is in agreement with its low RA value (~60%). However, the fracture morphologies for the P-added and P+Ce-added samples are fully ductile (Figure 2e,f), being well corresponding to their much higher RA values as shown in Figure 1. In addition, it is clear that the P+Ce-added sample exhibits bigger and deeper dimples than the P-added one, which is also well consistent with the difference between their RA values (P-added: ~70%; P+Ce-added: ~80%). At 850 °C, there are distinct differences in RA (Figure 1) and fracture morphology (Figure 2g–i) between the three steels. The unadded sample displays mainly an intergranular fracture characterized by intergranular facets accompanied with somewhat ductile characteristics, where the RA is only ~50%. However, the P-added and P+Ce-added samples still present ductile rupture. Compared with the P-added sample, the P+Ce-added one exhibits more visible and bigger dimples. This is well reflected in their RA values, being about 70% and 90%, respectively. When deformed at 1000 °C, all the samples exhibit a completely ductile fracture (Figure 2j–l), corresponding well to their high RA values (Figure 1). Consequently, the results of fracture characteristics further confirm that the combined addition of P and Ce has a positive effect on the HD behavior of SA508-4N RPV steel.

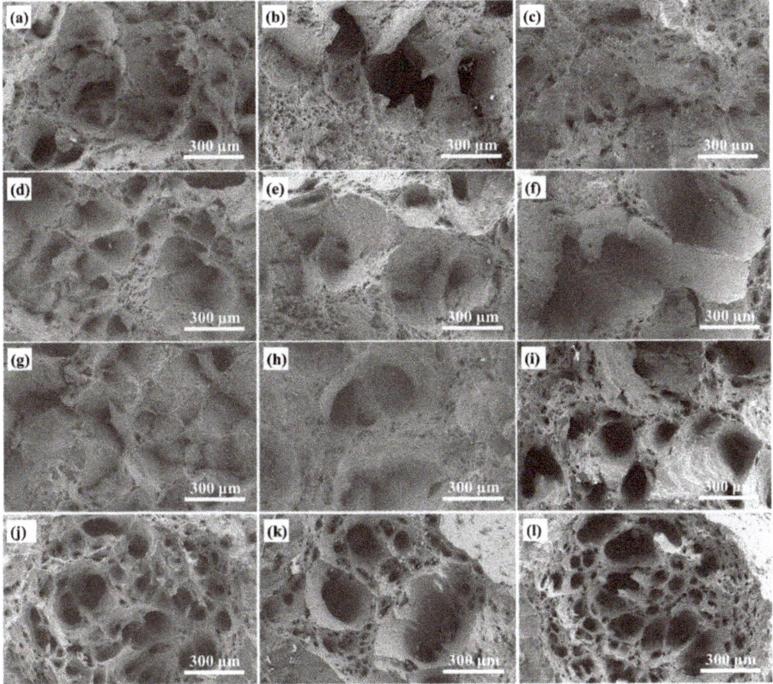

Figure 2. SEM fractographs of the specimens tested at different temperatures: (**a**) unadded steel at 750 °C; (**b**) P-added steel at 750 °C; (**c**) P+Ce-added steel at 750 °C; (**d**) unadded steel at 800 °C; (**e**) P-added steel at 800 °C; (**f**) P+Ce-added steel at 800 °C; (**g**) unadded steel at 850 °C; (**h**) P-added steel at 850 °C; (**i**) P+Ce-added steel at 850 °C; (**j**) unadded steel at 1000 °C; (**k**) P-added steel at 1000 °C; and (**l**) P+Ce-added steel at 1000 °C.

3.3. Microstructural Observation

The longitudinal microstructural observations near the fracture tip are presented in Figure 3. As shown, no visible proeutectoid ferrite is precipitated along prior austenite GBs for any type of the ruptured samples after the tensile test. This indicates that the HD behavior of the three steels is merely related to the deformation of austenite. If the DR and GB sliding do not take place during hot deformation, the deformation bands will be formed along the tensile direction. The finer the bands, the stronger the deformation ability, and thus the better is the HD behavior. When deformed at 750 °C, there is no obvious difference in the band-shaped structure between different steels (Figure 3a–c). This is in accord with their similar RA values (Figure 1). However, when deformed at 800 °C, there is an apparent difference in morphology (Figure 3d–f). The band-shaped structure becomes more evident from the unadded to P-added to P+Ce-added samples, implying that the deformation capability increases in the same order, as demonstrated in Figure 1. When deformed at 850 °C, the band-shaped structure is still seemingly visible in the unadded and P-added samples and more clear in the latter (Figure 3g,h), being coincident with their RA values (~50% and 70%, respectively). Nevertheless, there is no band-shaped structure formed in the P+Ce-added sample (Figure 3i). Owing to the fact that the P+Ce-added sample exhibits a very high RA at this temperature (~90%), the DR should have occurred during hot deformation. It is worth noting that the microstructural morphology of the unadded sample (Figure 3g) clearly indicates that some cracks are initiated and propagated along austenite grain boundaries, thereby causing intergranular fracture characterized by the sawtooth-shaped fracture tip. This is well in accordance with the fracture morphology as shown above. When deformed at 900 °C or above, there are no band-shaped structures in all deformed samples, but with a high HD performance (Figure 3j–l). Under these circumstances, it can be inferred that the DR has took place in the course of hot deformation at these temperatures. Therefore, the combined addition of trace P and Ce in the SA508-4N RPV steel may be able to lower the starting temperature of DR from ~900 to ~850 °C.

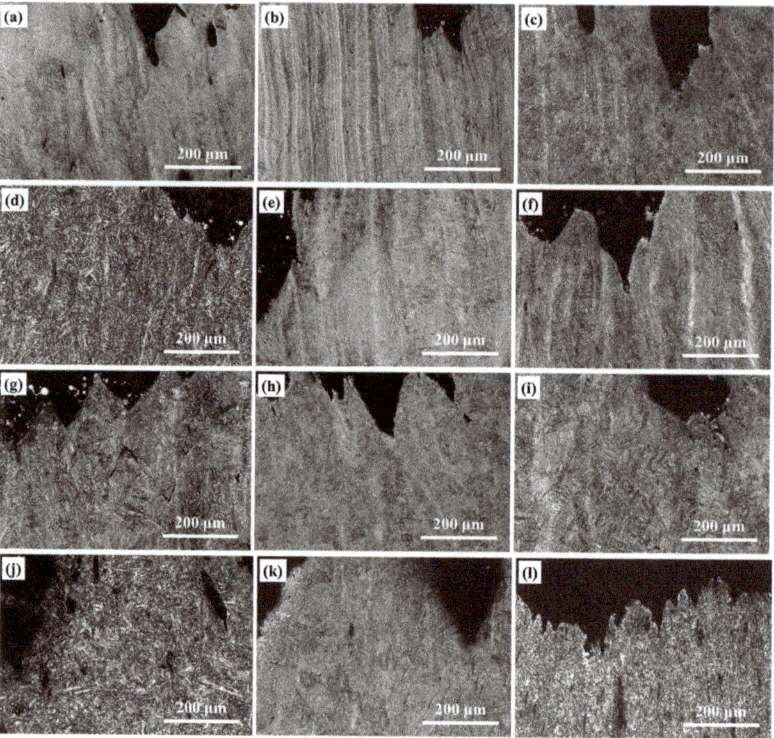

Figure 3. Optical microstructures near fracture tips of the specimens tested at different temperatures: (**a**) unadded steel at 750 °C; (**b**) P-added steel at 750 °C; (**c**) P+Ce-added steel at 750 °C; (**d**) unadded steel at 800 °C; (**e**) P-added steel at 800 °C; (**f**) P+Ce-added steel at 800 °C; (**g**) unadded steel at 850 °C; (**h**) P-added steel at 850 °C; (**i**) P+Ce-added steel at 850 °C; (**j**) unadded steel at 900 °C; (**k**) P-added steel at 900 °C; and (**l**) P+Ce-added steel at 900 °C.

4. Discussion

The following factors usually affect the HD of a steel. In the austenite–ferrite dual-phase region, the HD may be degenerated by a thin ferrite layer precipitated along the austenite GB since the hot deformation can be concentrated in this thin ferrite area, resulting in failure with a small overall deformation [9,13,14,34]. In the austenite single-phase region, the HD can be deteriorated by the GB sliding or substantially enhanced by the DR [9,12,14,35–38]. As the GB strength decreases with the increase of temperature, the GB sliding will become more serious at a higher temperature (i.e., the GB sliding increases with the increase of temperature), thereby reducing the HD of the steel. Generally, the GB sliding is affected by the GBS of solute or impurity atoms and the GB precipitation of carbides [15,39–41]. However, the DR would occur during hot deformation if the deformation temperature is sufficiently high, resulting in a dramatic enhancement of HD. Accordingly, the HD trough of a steel in the austenite single-phase range is caused by a combination of GB sliding and DR.

Besides, the austenite grain size is another factor influencing the HD of a steel. Normally, the samples with finer grains would exhibit higher RA values as compared with the coarse-grained samples.

Optical microscopy, as shown in Figure 3, indicates that all the three steels have no evident proeutectoid ferrite precipitation at prior austenite GBs after hot deformation, even at 750 °C. This implies that the A_{r3} of this RPV steel is lower than 750 °C; that is, the deformation between 750 and 1000 °C was proceeded in the austenite single-phase range. Since the microstructural characteristics are not very clear because of the deformation bands near the fracture tip as displayed in Figure 3, it is necessary to further confirm that the A_{r3} of the steel is definitely lower than 750 °C. For this, the samples were heated to 1300 °C for 180 s, then furnace-cooled to 750 °C and maintained there for 1 h, followed by water quenching. Evidently, the optical microstructure revealed by etching with nital, as shown in Figure 4a,c,e, is a typical quenched martensite with no proeutectoid ferrite formation. Consequently, the A_{r3} of the steel is lower than 750 °C indeed, which is well consistent with thermodynamic calculations where the A_3 (or A_{e3}) of this RPV steel is forecasted as ~750 °C [11,42,43]. To reveal the prior austenite grain boundaries, the samples were etched with a supersaturated picric acid aqueous solution. As shown in Figure 4b,d,f, the average austenite grain size has no noticeable difference between different steels, which are about 178, 191, and 186 μm for the unadded, P-added, and P+Ce-added steels, respectively, determined by the linear intercept method. Based on these results, it could be inferred that before hot deformation at each temperature, the average grain size of austenite had no apparent difference between different steels. Therefore, the difference in HD behavior between these steels may not be related to grain size. It is worth mentioning that since the HD measurements were all performed in the austenite single-phase region, the microstructural constituents after hot tensile testing and subsequent cooling are not concerned with the above HD behavior. Accordingly, it is unnecessary to take into account their details in the present research.

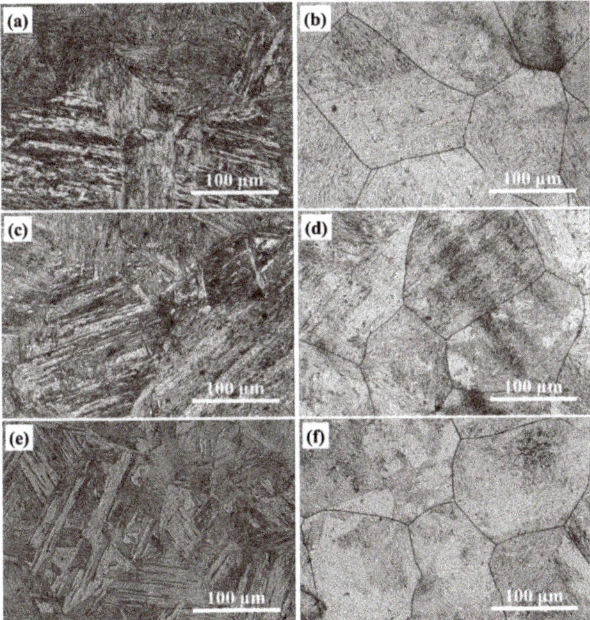

Figure 4. Optical micrographs of the specimens heated at 1300 °C for 180 s, furnace-cooled to 750 °C and held there for 1 h, followed by water quenching: (**a**,**b**) unadded steel; (**c**,**d**) P-added steel; (**e**,**f**) P+Ce-added steel; (**a**,**c**,**e**) etched with nital; (**b**,**d**,**f**) etched with supersaturated picric acid aqueous solution.

Many investigations [20,36,44] have demonstrated that the DR can relieve the concentrations of stress and dislocations so that the HD behavior can be significantly improved by restraining the microcrack formation. In other words, the starting temperature at which the DR may occur during hot deformation has a great impact on the HD performance of a steel. As inferred previously from Figures 1 and 3, the DR occurred during hot deformation at 850 °C for the P+Ce-added sample and at 900 °C or above for all the three types of samples. This will be further discussed tentatively on the basis of EBSD images. The EBSD images in the regions near the fracture tip of the samples tested at 800, 850, 900, and 1000 °C are illustrated in Figure 5. As observed in Figure 5a–c, there are the distinct oriented big blocks in all samples deformed at 800 °C, and within each big block, the colors for different small constituents have no large difference. With reference to Figures 1 and 3, it could be believed that the DR did not occur during hot deformation. It is also seen that the distinct oriented big blocks exist in both the unadded and P-added samples deformed at 850 °C (Figure 5d,e). For the P+Ce-added sample deformed at 850 °C, a mixture of oriented big blocks and equiaxed small blocks is mainly formed (Figure 5f), while the equiaxed small blocks present primarily in the other cases. This demonstrates that when deformed at 850 °C, the DR should have happened to some degree in the P+Ce-added sample, but it should have not in the unadded and P-added samples. When deformed at 900 °C or above, all EBSD images exhibit a similar morphology mainly with equiaxed small blocks. Accordingly, the DR should have fully occurred in these samples, leading to an excellent HD behavior as shown Figure 1.

The stress–strain curves at different temperatures are shown in Figure 6. It is clear that the tensile strength is generally higher for the P-added and P+Ce-added samples than for the unadded ones, and the peak stress shifts to higher strains. As is well acknowledged [45,46], to initiate DR during hot deformation at a certain temperature, a critical stored energy is required. Since 90%–95% of the deformation energy is normally dissipated by heat generation and only 5%–10% is stored in the defects within the sample, it is assumed that the stored energy in the material during deformation is about 10% of the deformation energy, and the uniform deformation can reach the peak stress. The deformation energy may be estimated by the area under the stress–strain curve before the peak stress, and thus the stored energy can be obtained. The stored energies are represented in Figure 7 for the samples deformed at 800–1000 °C. Apparently, when deformed at 850 °C, the stored energy increases on going from the unadded to P-added to P+Ce-added samples. It can be conceived that at 850 °C, the stored energies in the unadded and P-added samples cannot attain their corresponding critical stored energies for the initiation of DR, while the critical stored energy can be reached in the P+Ce-added samples. As a consequence, the DR in the P+Ce-added samples took place to some extent during hot deformation at 850 °C, raising its RA to a high level of ~90% (Figure 1). Nevertheless, this did not happen in the unadded and P-added samples. When deformed at 900 °C or above, the stored energies for all the samples should have well surpassed their corresponding critical energies, and hence the evident DR should have occurred in the course of hot deformation, thereby significantly enhancing the HD performance as illustrated in Figure 1.

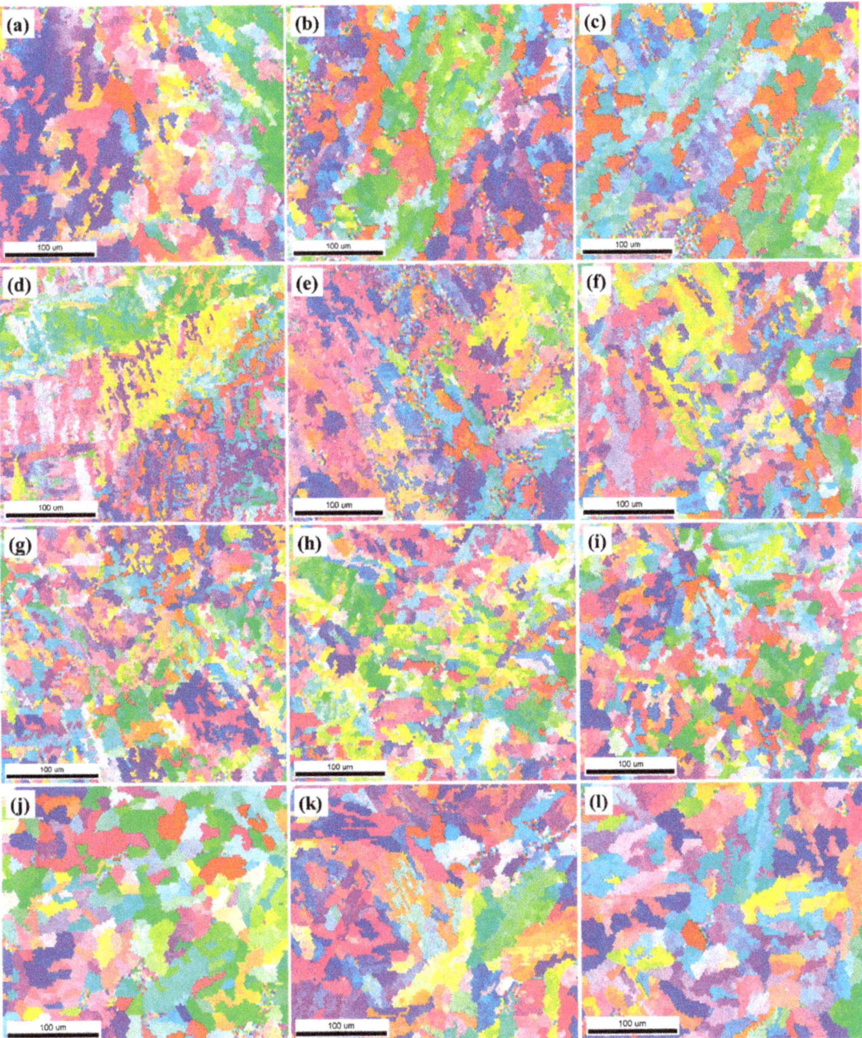

Figure 5. EBSD images showing microstructural features for the specimens tested at different temperatures: (**a**) unadded steel at 800 °C; (**b**) P-added steel at 800 °C; (**c**) P+Ce-added steel at 800 °C; (**d**) unadded steel at 850 °C; (**e**) P-added steel at 850 °C; (**f**) P+Ce-added steel at 850 °C; (**g**) unadded steel at 900 °C; (**h**) P-added steel at 900 °C; (**i**) P+Ce-added steel at 900 °C; (**j**) unadded steel at 1000 °C; (**k**) P-added steel at 1000 °C; and (**l**) P+Ce-added steel at 1000 °C.

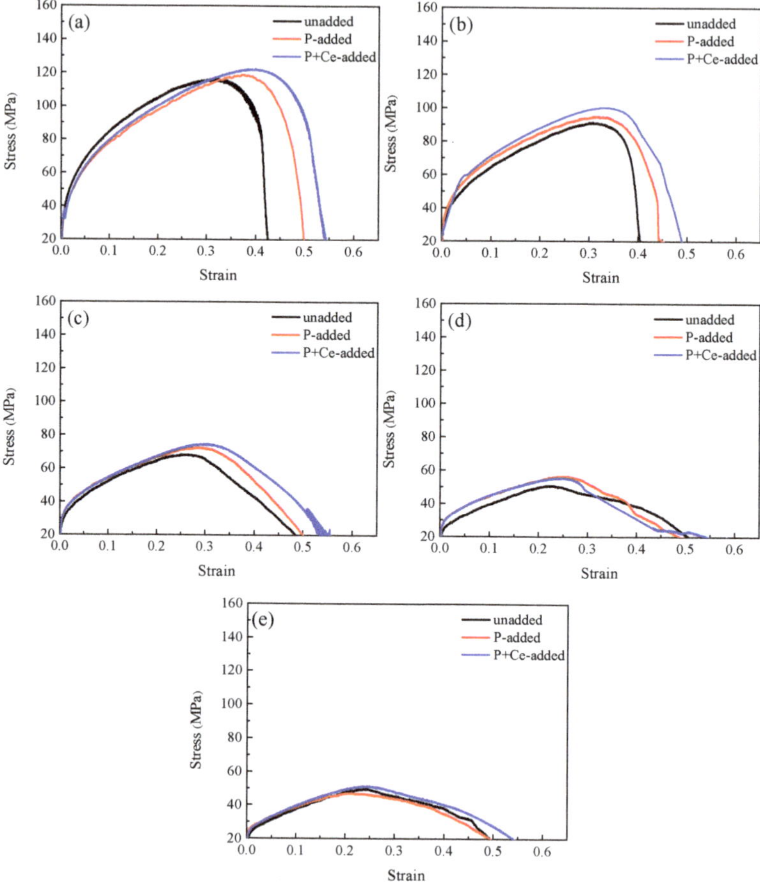

Figure 6. Stress–strain curves for the unadded, P-added, and P+Ce-added steels at different temperatures: (**a**) 800 °C, (**b**) 850 °C, (**c**) 900 °C, (**d**) 950 °C, and (**e**) 1000 °C.

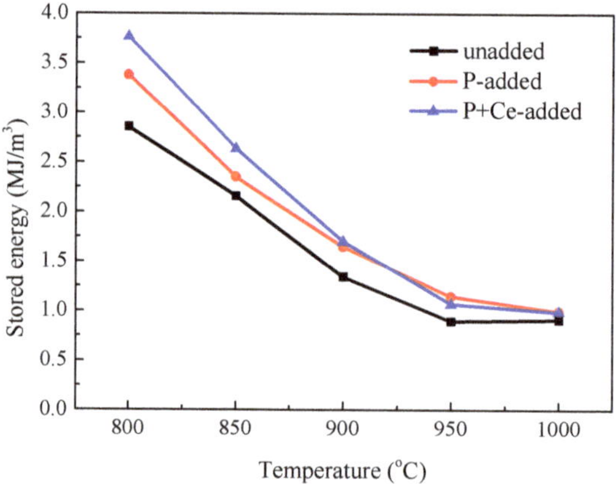

Figure 7. The stored energies in different samples deformed at different temperatures.

As clarified above, the trace addition of P and Ce can raise the HD of the steel in the case of no DR. This should be realized by restraining the GB sliding by their GBS. Since there is no occurrence of DR in the 800 °C-deformed samples and they have an apparent difference in RA, the boundary segregation of P and Ce in the P-added and P+Ce-added samples tested at this temperature was investigated using FE-STEM along with EDS, and the analyzed results are shown in Figure 8. As can be seen, the GB concentration of P is about 0.33 at.% for the P-added sample and 0.23 at.% for the P+Ce-added one, which are much higher than their respective matrix levels. The GB concentration of Ce is about 0.22 at.%, which is 18 times as high as its matrix level (approximately 0.012 at.%). Therefore, the GBS of P took place in the P-added and P+Ce-added samples, and that of Ce did in the P+Ce-added sample. As described in [11], the beneficial effect of P GBS on the HD of a steel might be explained as follows. The segregated P atoms could take up the boundary vacant sites like "cavities," thereby lessening the GB sliding as the boundary cavities could be more effective in facilitating the boundary sliding than the segregation of P. As for the effect of Ce, many studies [26–29,37] have revealed that the GBS of Ce can raise the GB cohesion. Due to the augmentation of boundary cohesion, the GB sliding can be suppressed to a certain degree, thus enhancing the HD performance of the steel. Moreover, the suppression of the GB sliding may enable the uniform hot deformation to extend a higher strain with a higher flow stress. This may make the deformed material possess more stored energy, thereby lowering the initial temperature of DR as shown above. In addition, Ce can suppress the GBS of P to a certain extent, reducing the detrimental effect of P on the boundary cohesion. Therefore, the GBS of Ce can further raise the RA value of the steel, from ~70% for the P-added sample to ~90% for the P+Ce-added one, completely eliminating the HD trough. These results are in agreement with previous studies [23,29,37,47]. The GBS of rare earth Ce can considerably improve the HD of a 1Cr-0.5Mo low-alloy steel by enhancing the boundary cohesion and lowering the initial temperature of DR and suppressing the deleterious effect of impurity Sn [23,37]. The segregation of Ce can also enhance the HD of a Sn-doped C-Mn RPV steel by the same mechanisms [29]. The GBS of rare earth La is able to improve the HD of a low-carbon steel doped with As and Sn by reducing the GBS of harmful impurities As and Sn and lowering the starting temperature of DR [47]. The GBS of solute atoms in steel is divided into equilibrium segregation (ES) and nonequilibrium segregation (NES). The ES is produced during equilibration of the system. It is a function of temperature and decreases exponentially with rising temperature [48,49]. The NES of solute atoms counts on the formation of supersaturated vacancy–solute complexes since there is a binding energy between vacancy and solute atom [50], and the vacancies are supersaturated in the course of cooling from a high heating temperature [51–53]. During cooling, the complexes diffuse from the grain interior to the boundary, causing the solute boundary segregation. Furthermore, the NES may also be produced during hot plastic deformation as the deformation can create excess vacancies [54,55].

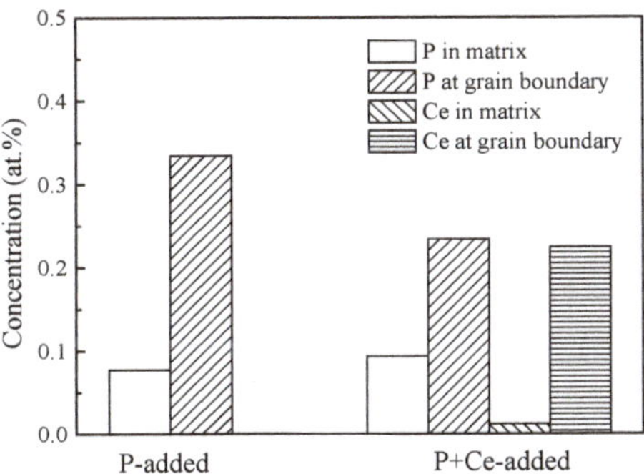

Figure 8. Grain boundary concentrations of P and Ce in the P-added and P+Ce-added specimens tested at 800 °C, determined by FE-STEM microanalysis.

As described above [11,15,39], carbide precipitation along the GBs also plays an important role in the HD behavior. As seen from TEM micrographs in Figure 9, there is no apparent carbide precipitation formed along the GBs when deformed at 800 °C for all the three types of samples. This consists well with the results of Thermo-Calc calculations, which indicate that no apparent $M_{23}C_6$-type carbides are precipitated above 750 °C in the SA508-4N steel [11,42]. Accordingly, the HD performance in the austenite single-phase range should have nothing to do with the GB precipitation of carbides for this RPV steel.

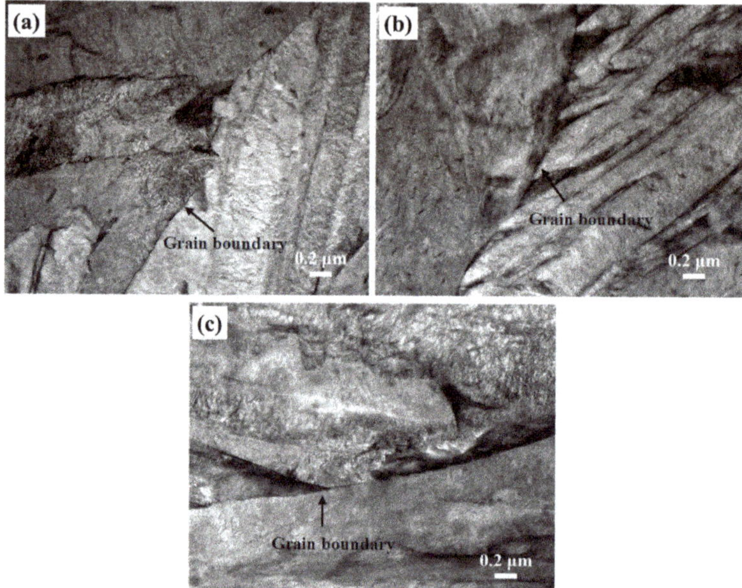

Figure 9. Transmission electron micrographs for the specimens tested at 800 °C: (**a**) unadded steel, (**b**) P-added steel, and (**c**) P+Ce-added.

5. Conclusions

The HD behaviors of three heats of SA508-4N RPV steel, unadded, P-added, and P+Ce-added, were investigated in detail over the range of 750–1000 °C. The testing temperatures are located in the austenite single-phase region. A deep HD trough exists for the unadded steel with the lowest RA value of ~50% at 850 °C, whereas there is no HD trough present for the P+Ce-added steel whose RA value is up to ~90% at this temperature, demonstrating that the trace addition of P and Ce can dramatically raise the HD of the steel, especially at temperatures between 800 and 900 °C. No doubt, the quality of continuous casting and hot working of the P+Ce-added steel can be improved considerably. The main mechanism for the above effect is suggested to be DR and GBS. The P+Ce addition in the steel may lower the initial temperature of DR from ~900 to ~850 °C. Meanwhile, the GBS of P and Ce may restrain the GB sliding by enhancing the GB cohesion. Consequently, the combined addition of minor P and Ce can significantly enhance the HD performance of SA508-4N RPV steel. No doubt, the highly enhanced HD performance of this steel can improve the quality and productivity of its continuous casting and hot working.

Author Contributions: Y.G.: investigation, analysis, and writing—original draft; Y.Z.: analysis and writing—review and editing; S.S.: conceptualization, supervision, and writing—review and editing. All authors have read and agreed to the published version of the manuscript.

Funding: This research was funded by the National Natural Science Foundation of China (Grant No. 51071060) and Natural Science Foundation of Anhui Province (Grant No. 2008085QE197).

Conflicts of Interest: The authors declare no conflict of interest.

References

1. Knott, J.F. Structural integrity of nuclear reactor pressure vessels. *Philos. Mag.* **2013**, *93*, 3835–3862. [CrossRef]
2. Zinkle, S.J.; Was, G.S. Materials challenges in nuclear energy. *Acta Mater.* **2013**, *61*, 735–758. [CrossRef]
3. Toribio, J.; Vergara, D.; Lorenzo, M. Role of in-service stress and strain fields on the hydrogen embrittlement of the pressure vessel constituent materials in a pressurized water reactor. *Eng. Fail. Anal.* **2017**, *82*, 458–465. [CrossRef]
4. Kim, M.C.; Park, S.G.; Lee, K.H.; Lee, B.S. Comparison of fracture properties in SA508 Gr.3 and Gr.4N high strength low alloy steels for advanced pressure vessel materials. *Int. J. Pres. Ves. Pip.* **2015**, *131*, 60–66. [CrossRef]
5. Lee, K.H.; Park, S.G.; Kim, M.C.; Lee, B.S.; Wee, D.M. Characterization of transition behavior in SA508 Gr.4N Ni-Cr-Mo low alloy steels with microstructural alteration by Ni and Cr contents. *Mater. Sci. Eng. A* **2011**, *529*, 156–163. [CrossRef]
6. Lee, K.H.; Park, S.G.; Kim, M.C.; Lee, B.S. Cleavage fracture toughness of tempered martensitic Ni-Cr-Mo low alloy steel with different martensite fraction. *Mater. Sci. Eng. A* **2012**, *534*, 75–82. [CrossRef]
7. Dai, X.; Yang, B. Hot deformation behavior and microstructural evolution of SA508-IV steel. *Steel Res. Int.* **2018**, *89*, 1800208. [CrossRef]
8. Thomas, B.G.; Samarasekera, I.V.; Brimacombe, J.K. Mathematical model of the thermal processing of steel ingots: Part II. Stress model. *Metall. Trans. B* **1987**, *18*, 131–147. [CrossRef]
9. Mintz, B.; Yue, S.; Jonas, J.J. Hot ductility of steels and its relationship to the problem of transverse cracking during continuous casting. *Metall. Rev.* **1991**, *36*, 187–217. [CrossRef]
10. Abushosha, R.; Vipond, R.; Mintz, B. Influence of sulphur and niobium on hot ductility of as cast steels. *Mater. Sci. Technol.* **1991**, *7*, 1101–1107. [CrossRef]
11. Guo, Y.; Wang, K.; Song, S.H. Abnormal influence of impurity element phosphorus on the hot ductility of SA508Gr.4N reactor pressure vessel steel. *Mater. Sci. Eng. A* **2020**, *79*, 139837. [CrossRef]
12. Mintz, B.; Cowley, A.; Talian, C.; Crowther, D.N.; Abushosha, R. Influence of P on hot ductility of high C, Al, and Nb containing steels. *Mater. Sci. Technol.* **2003**, *19*, 184–188. [CrossRef]
13. Jiang, X.; Chen, X.M.; Song, S.H.; Shangguan, Y.J. Phosphorus-induced hot ductility enhancement of 1Cr–0.5Mo low alloy steel. *Mater. Sci. Eng. A* **2013**, *574*, 46–53. [CrossRef]
14. Mintz, B.; Arrowsmith, J.M. Hot-ductility behaviour of C-Mn-Nb-Al steels and its relationship to crack propagation during the straightening of continuously cast strand. *Met. Technol.* **1979**, *6*, 24–32. [CrossRef]
15. Maehara, Y.; Ohmori, Y. The precipitation of AlN and NbC and the hot ductility of low carbon steels. *Mat. Sci. Eng.* **1984**, *62*, 109–119. [CrossRef]
16. Kang, S.E.; Banerjee, J.R.; Tuling, A.; Mintz, B. Influence of P and N on hot ductility of high Al, boron containing TWIP steels. *Mater. Sci. Technol.* **2014**, *30*, 1328–1335. [CrossRef]
17. Zheng, L.; Song, S.H. Effect of thermal cycling induced phosphorus grain boundary segregation on embrittlement of welding heat affected zones in 2.25Cr-1Mo steel. *Mater. Sci. Technol.* **2014**, *30*, 1378–1385.
18. Chen, X.M.; Song, S.H.; Weng, L.Q.; Liu, S.J.; Wang, K. Relation of ductile-to-brittle transition temperature to phosphorus grain boundary segregation for a Ti-stabilized interstitial free steel. *Mater. Sci. Eng. A* **2011**, *528*, 8299–9304. [CrossRef]
19. Misra, R.D.K. Temperature-time dependence of grain boundary segregation of phosphorus in interstitial-free steels. *J. Mater. Sci. Lett.* **2002**, *21*, 1275–1279. [CrossRef]
20. Yu, Y.C.; Chen, W.Q.; Zheng, H.G. Effects of Ti-Ce refiners on solidification structure and hot ductility of Fe-36Ni invar alloy. *J. Rare Earth.* **2013**, *31*, 927–932. [CrossRef]
21. Liu, H.L.; Liu, C.J.; Jiang, M.F. Effect of rare earths on impact toughness of a low-carbon steel. *Mater. Design* **2012**, *33*, 306–312. [CrossRef]
22. Song, S.H.; Sun, H.J.; Wang, M. Effect of rare earth cerium on brittleness of simulated welding heat-affected zones in a reactor pressure vessel steel. *J. Rare Earth.* **2015**, *33*, 1204–1211. [CrossRef]

23. Jiang, X.; Song, S.H. Enhanced hot ductility of a Cr–Mo low alloy steel by rare earth cerium. *Mater. Sci. Eng. A* **2014**, *613*, 171–177. [CrossRef]

24. Ma, X.C.; An, Z.J.; Chen, L.; Mao, T.Q.; Wang, J.F.; Long, H.J.; Xue, H.Y. The effect of rare earth alloying on the hot workability of duplex stainless steel-A study using processing map. *Mater. Des.* **2015**, *86*, 848–854. [CrossRef]

25. Guo, Y.; Sun, S.F.; Song, S.H. Effect of minor rare earth cerium addition on the hot ductility of a reactor pressure vessel steel. *Results Phys.* **2019**, *15*, 102746. [CrossRef]

26. Xin, W.B.; Zhang, J.; Luo, G.P.; Wang, R.; Meng, Q.Y.; Song, B. Improvement of hot ductility of C-Mn Steel containing arsenic by rare earth Ce. *Metall. Res. Technol.* **2018**, *115*, 419. [CrossRef]

27. Wang, H.Y.; Gao, X.Y.; Ren, H.P.; Zhang, H.W.; Tan, H.J. First-principles characterization of lanthanum occupying tendency in α-Fe and effect on grain boundaries. *Acta Phys. Sin-Ch. Ed.* **2014**, *63*, 148101.

28. Liu, G.L.; Zhang, G.Y.; Li, R.D. Electronic Theoretical Study of the Interaction between Rare Earth Elements and Impurities at Grain Boundaries in Steel. *J. Rare Earth.* **2003**, *21*, 372–374.

29. Song, S.H.; Sun, S.F. Effect of Rare-Earth Cerium on Impurity Tin-Induced Hot Ductility Deterioration of SA508-III Reactor Pressure Vessel Steel. *Steel Res. Int.* **2016**, *87*, 1435–1443. [CrossRef]

30. Hannerz, N.E. Critical hot plasticity and transverse cracking in continuous slab casting with particular reference to composition. *ISIJ. Int.* **1985**, *25*, 149–158. [CrossRef]

31. Mintz, B.; Abushosha, R. Effectiveness of hot tensile test in simulating straightening in continuous-casting. *Mater. Sci. Technol.* **1992**, *8*, 171–177. [CrossRef]

32. Song, S.H.; Weng, L.Q. An FEGSTEM study of grain boundary segregation of phosphorus during q-uenching in a 2.25Cr-1Mo steel. *J. Mater. Sci. Technol.* **2005**, *21*, 445–450.

33. Chen, X.M.; Song, S.H.; Weng, L.Q.; Liu, S.J. Solute grain boundary segregation during high temperature plastic deformation in a Cr-Mo low alloy steel. *Mater. Sci. Eng. A* **2011**, *528*, 7663–7668. [CrossRef]

34. Mintz, B.; Abushosha, R.; Shaker, M. Influence of deformation induced ferrite, grain boundary sliding, and dynamic recrystallisation on hot ductility of 0.1–0.75% C steels. *Mater. Sci. Technol.* **1993**, *9*, 907–914. [CrossRef]

35. Mintz, B.; Cowley, A. Deformation induced ferrite and its influence on the elevated temperature tensile flow stress-elongation curves of plain C-Mn and Nb containing steels. *Mater. Sci. Technol.* **2006**, *22*, 279–292. [CrossRef]

36. Mintz, B. The influence of composition on the hot ductility of steels and to the problem of transverse cracking. *ISIJ. Int.* **1999**, *39*, 833–855. [CrossRef]

37. Song, S.H.; Xu, Y.W.; Chen, X.M.; Jiang, X. Effect of rare earth cerium and impurity tin on the hot ductility of a Cr-Mo low alloy steel. *J. Rare Earth.* **2016**, *34*, 1062–1068. [CrossRef]

38. Wang, K.; Yang, C.; Si, H. Effect of non-equilibrium grain-boundary segregation of phosphorus on the intermediate-temperature ductility minimum of an Fe-17Cr alloy. *Mater. Sci. Eng. A* **2013**, *587*, 228–232. [CrossRef]

39. Ouchi, C.; Matsumoto, K. Hot ductility in Nb-bearing high-strength low-alloy steels. *ISIJ. Int.* **2006**, *22*, 181–189. [CrossRef]

40. Lv, B.; Zhang, F.C.; Li, M.; Hou, R.J.; Qian, L.H.; Wang, T.S. Effects of phosphorus and sulfur on the thermoplasticity of high manganese austenitic steel. *Mater. Sci. Eng. A* **2010**, *527*, 5648–5653. [CrossRef]

41. Song, S.H.; Jiang, X.; Chen, X.M. Tin-induced hot ductility degradation and its suppression by phosphorus for a Cr-Mo low-alloy steel. *Mater. Sci. Eng. A* **2014**, *595*, 188–195. [CrossRef]

42. Li, C.Y.; Liu, Z.D.; Lin, Z.J. Thermodynamic calculation and analysis on equilibrium phase in low alloy steel for reactor pressure vessel. *Heat. Treat. Met.* **2013**, *38*, 14–17.

43. Liu, N.; Liu, Z.D.; He, X.K.; Yang, Z.Q. Austenite transformation of SA508Gr.4N steel for nuclear reactor pressure vessels during heating process. *Heat. Treat. Met.* **2017**, *3*, 88–92.

44. Mintz, B.; Cowley, A.; Abushosha, R.; Crowther, D.N. Hot ductility curve of an austenitic stainless steel and importance of dynamic recrystallisation in determining ductility recovery at high temperatures. *Mater. Sci. Technol.* **1999**, *15*, 1179–1185. [CrossRef]

45. Manda, D.; Baker, I. Determination of the stored energy and recrystallization temperature as a function of depth after rolling of polycrystalline copper. *Scripta Mater.* **1995**, *33*, 645–650. [CrossRef]

46. Zhang, Y.; Wang, J.T.; Cheng, C.; Liu, J.Q. Stored energy and recrystallization temperature in high purity copper after equal channel angular pressing. *J. Mater. Sci.* **2008**, *43*, 7326–7330. [CrossRef]

47. Huang, C.G.; Song, B.; Xin, W.B.; Jia, S.J.; Yang, Y.H. Influence of rare earth La on hot ductility of low carbon steel containing As and Sn. *Heat Treat. Met.* **2015**, *40*, 1–6.

48. McLean, D. *Grain Boundaries in Metals*; Oxford University Press: London, UK, 1957; pp. 116–148.

49. Lejcek, P. *Grain Boundary Segregation in Metals*; Springer: Berlin, Germany, 2010; pp. 50–102.

50. Faulkner, R.G.; Song, S.H.; Flewitt, P.E.J. Determination of impurity-point defect binding energies in alloys. *Mater. Sci. Technol.* **1996**, *12*, 904–910. [CrossRef]

51. Faulkner, R.G. Nonequilibrium grain boundary segregation in austenitic alloys. *J. Mater. Sci.* **1981**, *16*, 373–383. [CrossRef]

52. Song, S.H.; Faulkner, R.G.; Flewitt, P.E.J. Effect of boron on phosphorus-induced temper embrittlement. *J. Mater. Sci.* **1999**, *34*, 5549–5556. [CrossRef]

53. Xu, T.D.; Cheng, B.Y. Kinetics of non-equilibrium grain-boundary segregation. *Prog. Mater. Sci.* **2004**, *49*, 109–208. [CrossRef]

54. Song, S.H.; Zhang, Q.; Weng, L.Q. Deformation-induced non-equilibrium grain boundary segregation in dilute alloys. *Mater. Sci. Eng. A* **2008**, *473*, 226–232. [CrossRef]

55. Chen, X.M.; Song, S.H.; Weng, L.Q.; Wang, K. A study of deformation-induced phosphorus grain boundary segregation in an interstitial free steel. *Mater. Sci. Eng. A* **2012**, *545*, 86–90. [CrossRef]

Publisher's Note: MDPI stays neutral with regard to jurisdictional claims in published maps and institutional affiliations.

Article

Phase Field Simulation of AA6XXX Aluminium Alloys Heat Treatment

Antonis Baganis [1,*], Marianthi Bouzouni [1,2,*] and Spyros Papaefthymiou [1,*]

[1] Laboratory of Physical Metallurgy, Division of Metallurgy and Materials, School of Mining and Metallurgical Engineering, 9, Her. Polytechniou Str., Zografos, 15780 Athens, Greece

[2] Department of Physical Metallurgy and Forming, Hellenic Research Centre for Metals (ELKEME S.A.), 61st km Athens-Lamia Nat. Road, Oinofyta, 32011 Viotia, Greece

* Correspondence: mm15047@central.ntua.gr (A.B.); mbouzouni@elkeme.vionet.gr (M.B.); spapaef@metal.ntua.gr (S.P.); Tel.: +30-694-637-4689 (A.B.); +30-698-064-2333 (M.B.); +30-210-772-4710 (S.P.)

Abstract: Heat treatment has a significant impact on the microstructure and the mechanical properties of Al-Mg-Si alloys. The present study presents a first Phase-Field modelling approach on the recrystallisation and grain growth mechanism during annealing. It focuses on the precipitate fraction, radius, and Mg-Si concentration in the matrix phase, which are used as input data for the calculation of the yield strength and hardness at the end of different ageing treatments. Annealing and artificial ageing simulations have been conducted on the MultiPhase-Field based MICRESS@ software, while the ThermoCalc@ software has been used to construct the pseudo-binary Al-Mg phase-diagrams and the atomic-mobility databases of Mg_xSi_y precipitates. Recrystallisation simulation estimates the recrystallisation kinetics, the grain growth, and the interface mobility with the presence/absence of secondary particles, selecting as annealing temperature 400 °C and a microstructure previously subjected to cold rolling. The pinning force of secondary particles decelerates the overall recrystallisation time, causing a slight decrease in the final grain radius due to the reduction of interface mobility. The ageing simulation examines different ageing temperatures (180 and 200 °C) for two distinct ternary systems (Al-0.9Mg-0.6Si/Al-1.0Mg-1.1Si wt.%) considering the interface energy and the chemical free energy as the driving force for precipitation. The combination of Phase-Field and the Deschamps–Brechet model predicted the under-ageing condition for the 180 °C ageing treatment and the peak-ageing condition for the 200 °C ageing treatment.

Keywords: Al-Mg-Si alloys; phase-field; heat-treatment; recrystallisation; ageing; precipitation hardening; micress; thermocalc

---

Citation: Baganis, A.; Bouzouni, M.; Papaefthymiou, S. Phase Field Simulation of AA6XXX Aluminium Alloys Heat Treatment. *Metals* **2021**, *11*, 241. https://doi.org/10.3390/met11020241

Academic Editor: Elisabetta Gariboldi
Received: 2 January 2021
Accepted: 26 January 2021
Published: 1 February 2021

Publisher's Note: MDPI stays neutral with regard to jurisdictional claims in published maps and institutional affiliations.

1. Introduction

Nowadays, the demand to reduce the overall automobile weight as a prerequisite for improved fuel efficiency has increased the interest for the lightweight aluminium alloys, which combine a high strength to weight ratio, an excellent response to mechanical and chemical corrosion, and superb formability and weldability [1]. Among the different series of aluminium alloys, vital is the role of the heat-treated 6XXX alloys, with Mg and Si as the main alloying elements, which are used in a wide range of technological sectors, including the automobile for automotive skins, the naval industry for marine vessels, and the aerospace industry for missiles, as well as extrusion profiles and pipelines [1–3]. The conventional processing of AA6XXX aluminium alloys consists of a thermomechanical treatment, usually rolling or extrusion, and a subsequent heat treatment including annealing and isothermal artificial ageing. The annealing treatment includes successive and competitive stages of recovery, static recrystallisation, and grain growth resulting in microstructures consisting of un-deformed equiaxed fine grains. The difference in stored volume energy between the deformed non-recrystallized and undeformed recrystal-

49

lized grains acts as a driving force for the annealing, while nanoparticles, primarily iron intermetallic particles, retard recrystallisation through the Zener pinning action [4].

The artificial ageing leads to the successive precipitation of Mg_xSi_y intermetallic particles, which improve the yield strength and hardness of Al-Mg-Si alloys impeding the dislocation movement by shearing or bypassing (Orowan Loop) mechanism [1].

The optimization of AA6XXX alloys processing requires the development of fast, precise and versatile modelling techniques for both the annealing and the isothermal artificial ageing treatment. For both processes, the semi-empirical JMAK equations, which rely on randomness, have extensively been used [5]. However, Lan et al. [6] and Jou et al. [7] have shown that the Phase-Field modelling more accurately approached the analytical solution of recrystallisation in comparison to the JMAK model. For the isothermal artificial ageing process, numerous models have been proposed, considering the alloying elements concentration and the ageing conditions (time and temperature) in order to predict the precipitation volume fraction and radius, and their contribution to the overall strengthening of the alloy. These models lie on both the atomistic scale, as the Kinetic Monte Carlo approach [8], and the mesoscale, like the Kampman–Wagner [9] and the KiNG model [10], while the thermodynamic approach of CALPHAD has also been taken into account [11].

One of the most prominent numerical approaches for the evolution of precipitation mechanism is the mesoscale based MultiPhase-Field modelling, which incorporates the thermodynamic equations of Gibbs–Thompson that describe the dissolution/growth—of precipitates—and the Onsager equations, which describe the dissipation of free energy [12], while it can be coupled with the CALPHAD approach. In the MultiPhase-Field approach, microstructure is represented continuously using an order parameter φ, as a function of position and time, with a specific order parameter attributed to every single phase or even every grain of the system taking the value 0 for the matrix and the value 1 for precipitates. The interfaces are set as narrow regions, whose phase-field variables vary gradually between their values in the neighbouring grains, in the range $0 < \varphi < 1$, so that the model is described as a diffuse-interface approach, which secures an accurate and computationally flexible description of interfaces and their importance for the precipitation phenomena. The MultiPhase-Field models can insert, concurrently, the reduction of chemical free energy, the interface energy, and the elastic energy as driving force for the evolution of microstructure during ageing, while they can beneficially deal with phenomena, which are characterized by the overlap between the diffusion fields of particles which grow from different locations [13,14]. Phase-Field modelling has been used for various phase transformation simulations including solidification [15], martensite transformation [16], the $\alpha \rightarrow \gamma$ transformation [17], and the ageing of nickel superalloys [18,19] and 2XXX aluminium alloys [20]. The use of Phase-Field approach for numerical simulation and microstructure evolution of the aged Al-Mg-Si alloys is an innovative and demanding procedure considering the absence of literature and the time consumable simulation.

The current work is divided into the annealing and the ageing simulation. The annealing simulation focuses on the recrystallisation and grain growth mechanism, identifying the impact of secondary particles-pinning force on the recrystallisation kinetics and the average radius of recrystallized grains. The ageing simulation examines the evolution of the volume fraction and the radius of precipitates for varying ageing conditions and chemical compositions, while estimating the yield strength and hardness by the end of every single ageing simulation.

2. Materials and Methods

2.1. Al-Mg Phase Diagrams

The row of the simulation's steps is depicted in Figure 1. Al-Mg pseudo-binary phase diagrams are constructed with constant Si wt.% with the use of the ThermoCalc[®®] software (version 2019b, Royal Institute of Technology-KTH, Stockholm, Sweden) based on the TCAL4 thermodynamic database of aluminium alloys. Based on literature [2,4], two distinct chemical compositions are selected, Table 1, attributed to AA6061-AA6063 (1st case) and AA6082 (2nd case), respectively. The constructed phase diagrams are illustrated in Figure 2.

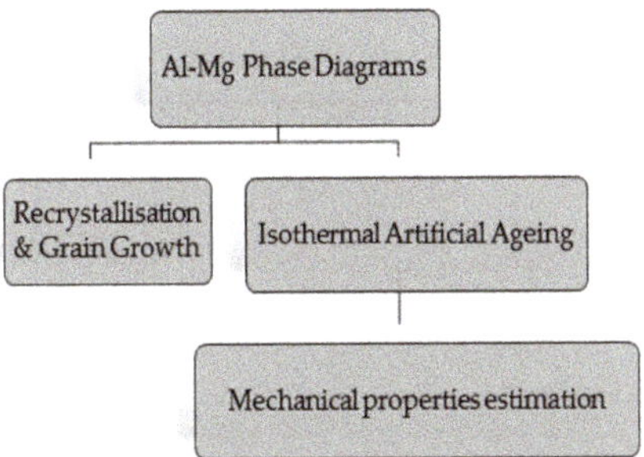

Figure 1. Simulation procedure analysis.

Table 1. Selected compositions of AA6XXX aluminium alloys [1,2,4].

Element	Mg	Si	Fe	Mn	Al
1st case (wt.%)	0.9	0.6	0.35	0.1	98.05 (bal.)
2nd case (wt.%)	1.0	1.1	0.35	0.1	97.45 (bal.)

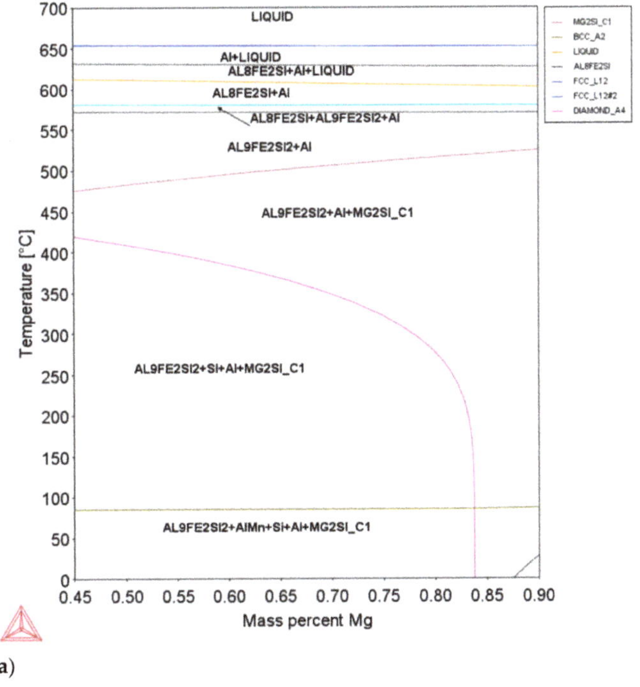

(a)

Figure 2. *Cont.*

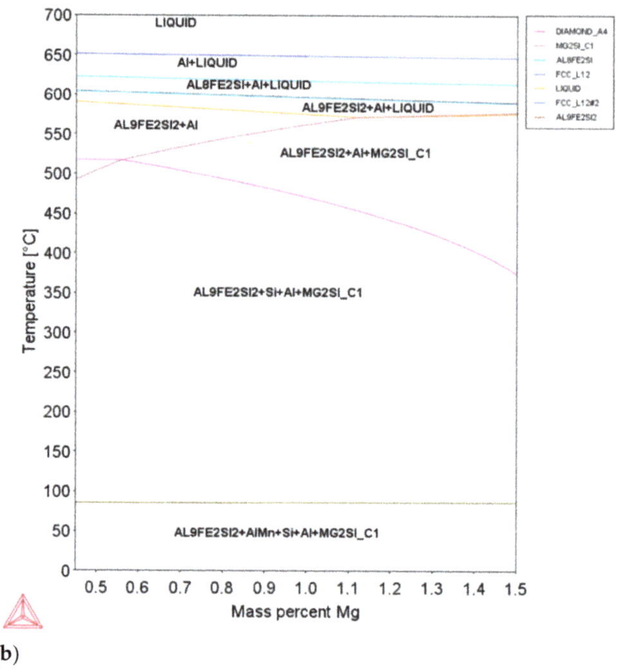

(b)

Figure 2. Equilibrium pseudo-binary Al-Mg phase diagrams, with constant Si concentration equal to 0.6 wt.% for (**a**) and 1.1 wt.% for (**b**).

2.2. Phase-Field Modelling

For the simulation of Al-Mg-Si alloys heat treatment, the MultiPhase-Field based MICRESS®® software (version 6.402) has been used, a computational tool developed by ACCESS e.V. research centre at the RWTH Aachen University in Germany. The equation of the phase-field variable (φ) of the recrystallisation mechanism is given in Equation (1), where τ is the kinetic parameter, e is the energy gradient coefficient, $g(\varphi)$ is a double-well function with heigh h, $p(\varphi)$ is the interpolation function, while the difference between the free energy of the deformed and the undeformed-recrystallized microstructure ($f_{Def} - f_{Rex}$) acts as the driving force for the recrystallisation mechanism [6]. The evolution of precipitate phase order parameter ($\dot{\varphi}_p$) is given in Equation (2), based on the modification of Steinbach's proposal [21] for the solid–liquid systems, where μ_{pm}, σ_{pm}, l_{pm}, ΔG_{pm} describe the interface mobility, interface energy, interface width, and the driving force, respectively, while φ_m is the order parameter of the matrix phase.

$$\frac{\partial \varphi}{\partial t} = \left(\frac{1}{\tau}\right)[e^2\nabla^2\varphi - h\left(\frac{\partial g}{\partial \varphi}\right) + \partial p/\partial \varphi\left(f_{Def} - f_{Rex}\right)] \tag{1}$$

$$\dot{\varphi}_p = \mu_{pm}\left[\sigma_{pm}\left\{\nabla^2\varphi_p - \left(\frac{\varphi_p\varphi_m}{l_{pm}^2}\right)(\varphi_p - \varphi_m)\right\} + \Delta G_{pm}\left(\frac{1}{l_{pm}}\right)\varphi_p\varphi_m\right] \tag{2}$$

2.3. Annealing Treatment Simulation

For the annealing treatment simulation, through MICRESS®® software, a suitable file has been created including the necessary parameters (Table 2) that describe the recrystallisation and grain growth mechanism, while the recovery mechanism is not taken into account. The estimation of the energy threshold (ΔE), as a driving force of recrystallisation, is given in Equation (3), where μ is the shear modulus, b the burger vector, and p_i the dislocation density, attributed to cold rolling of Al-Mg-Si alloys [22]. The simulated microstructure consists only of the matrix aluminium phase, whereas the impact of the secondary iron phase and Mg_xSi_y nanoparticles is taken into consideration via their pinning action. Cube texture [22,23] appears to be the most important crystallographic component of Al-Mg-Si alloys rolling and extrusion, and consequently its respective Miller indices have been inserted to the deformed microstructure. For the avoidance of a possible overlapping among the initial grains, the Voronoi tessellation has been enacted. Seeds of the recrystallized grains are set to nucleate on both the interior of the deformed grains and the interfaces. Two simulations have been conducted: One with the presence of pinning action, and one without the pinning action.

$$\Delta E = 0.5\mu b^2 p_i \tag{3}$$

Table 2. Principal parameters of annealing simulation.

Parameter	Value
Annealing Conditions	400 °C/5 min [22]
Interface Energy	0.32 J/m^2 [22]
Interface Mobility	3.8×10^{-5} cm^4/Js [24]
Shear Modulus Equation	$(84.8{-}4.06 \times 10^{-2} \times T)/(2 \times (1+v))$ [22]
Poisson Ration (v)	0.33 [22]
Burger Vector	0.286 nm [22]
Dislocation Density (p_i)	0.5×10^{14} 1/m^2 [22]
Energy Threshold	4.42×10^{-2} MPa
Pinning Force	0.18 1/μm [25]
Miller Indices (hkl) and (uvw)	001 and 100 respectively
Number of Cells ($x \times z$)	500×500
Cell Dimension	0.5 μm
Maximum Rotation Angle	15°
Prefactor of Interfacial Energy for Low-angle Grain Boundaries	0.2
Prefactor of Interfacial Mobility for Low-angle Grain Boundaries	0.1

2.4. Isothermal Artificial Ageing Simulation

For the case of isothermal artificial ageing simulation, mesh analysis is conducted, considering the importance of high resolution in depicting the nanoscopic size of precipitation particles and in calculating their critical curvature. The main parameters of mesh analysis are given in Table 3. For both the mesh analysis and the subsequent ageing simulations, the overall microstructure has been set equal to 20×20 μm^2. In order to insert a sufficient number of matrix phase grains, they are set to have a quite low average diameter, between 7.7 and 10 μm, which lies in the lower price range [26–28], while the ageing temperature does not permit recrystallisation to take place. It is worth noting that the applied values of the Al matrix grain diameter are not the same as the grain size estimated in the recrystallisation simulation. The reason for this variation is the necessity to insert a quite large number of Al grains (13) in the overall microstructure so as to be able to have a representative depiction of the spatial distribution of precipitate particles. As in the case of annealing, the Voronoi tessellation has been used for the improvement of the initial microstructure. An equal and quite large number of seeds are set to precipitate on both the interior of the matrix phase grains and the interfaces (3.75 particles/μm^2), while a shield distance (1 μm) and shield time have been set (1 s) for the nucleation and the initial growth or precipitates. It is worth noting, that the incubation time is considered to be negligible. For the ageing simulation, the ternary Al-Mg-Si system is used, as Mg and Si are the alloying elements which participate in the β-series particles (Mg_xSi_y).

Table 3. Principal parameters of mesh analysis.

Parameter	Value
Ageing Conditions	220 °C-4 h [29]
Number of Cells ($x \times z$)	500/800/1100/1400/1700/2000
Cell Dimension (nm)	40/25/18.2/14.3/11.8/10
Composition wt.%	Al-0.9Mg-0.6Si

Using ThermoCalc$^{®®}$, the necessary thermodynamic (TCAL4) and mobility (MOBAL3) databases of Mg_xSi_y phase have been constructed and inserted to MICRESS file, through TQ interface. Precipitate particles are considered to have spherical morphology, with constant chemical formula (Mg_2Si). The main parameters of ageing simulations are given in Table 4. The mobility pre-factors were adjusted, after trial-and-error sequences, to stabilize the matrix phase grains. The prices of molar volumes were taken from ThermoCalc$^{@}$. Given that precipitates are assumed to have spherical morphology, the elastic driving force has not been considered, as according to studies [30] it appears to have insignificant effect on the overall volume fraction of precipitates. Overall, 4 ageing simulations have been conducted, considering the ternary systems Al-0.9Mg-0.6Si and Al-1.0Mg-1.1Si (wt.%) for each of the following conditions: 180 °C-8 h [1,5,31], 200 °C-8 h [1,32].

Table 4. Principal parameters of ageing simulation.

Parameter	Value
Microstructure dimension	$20 \times 20 \ \mu m^2$
Interface energy (Al phase/Al phase-Mg_2Si)	$0.26 \ J/m^2$ [1]/$0.18 \ J/m^2$ [33]
Interface mobility (Al phase/Al phase-Mg_2Si)	$3.2 \times 10^{-13} \ m^4/Js$/$2 \times 10^{-13} \ m^4/Js$
Matrix Phase molar volume	$10.1 \times 10^{-6} \ m^3/mol$
Mg_2Si Phase molar volume	$12.9 \times 10^{-6} \ m^3/mol$

2.5. Yield Strength and Hardness estimation

The outputs of MultiPhase-Field analysis are used so as to estimate the yield strength and hardness, calculated via Equations (4) and (9), respectively [9,34]. The value of yield strength (σ_{ys}) is calculated as the sum of the grain-boundary strengthening ($\Delta\sigma_{gb}$), the solid solution strengthening ($\Delta\sigma_{ss}$), the modulus strengthening ($\Delta\tau_{ms}$), and the precipitation strengthening ($\Delta\tau_{ppt}$), based on Deschamps–Brechet model [34]. In Equation (5), k^j and c_i^j are the scaling factor and the weight % concentration, respectively, of j solute elements, where j = Mg and Si. In Equations (6)–(8), r is the radius and f the volume fraction of precipitates, b is a parameter equal to 0.5, while r_c is the transition radius from shearing (Equation (6)) to bypassing mechanism (Equation (8)). The term G expresses the shear modulus of Al matrix phase, whereas the term ΔG describes the difference of shear modulus between the matrix phase and the precipitates. The basic parameters for the yield strength and hardness calculation are given in Table 5.

$$\sigma_{ys} = \Delta\sigma_{gb} + \Delta\sigma_{ss} + M\sqrt{\Delta\tau_{ms}^2 + \Delta\tau_{ppt}^2} \tag{4}$$

$$\Delta\sigma_{ss} = \sum_j k^j C_i^{j2/3} \tag{5}$$

$$\sigma_{ppt} = \frac{1}{br}\left(2\beta Gb^2\right)^{-\frac{1}{2}}\left(\frac{3f}{2\pi}\right)^{1/2}\left[\left(2\beta Gb^2\right)\left(\frac{r}{r_c}\right)\right]^{3/2} \tag{6}$$

$$\Delta\tau_{ms} = 0.9\sqrt{(rf)}\left(\frac{\Gamma}{b}\right)\left(\frac{\Delta G}{G}\right)^{3/2}\left(2bln\left(\frac{2r}{b\sqrt{f}}\right)\right)^{-3/2} \tag{7}$$

$$\sigma_{ppt} = \frac{1}{br}\left(2\beta Gb^2\right)^{-\frac{1}{2}}\left(\frac{3f}{2\pi}\right)^{\frac{1}{2}}\left(2\beta Gb^2\right)^{3/2} \tag{8}$$

$$HV = 0.33\sigma_y + 16 \tag{9}$$

Table 5. Input parameters for yield strength and hardness calculation.

Parameter	Value
k_{Si}	66.3 MPa/wt%^{3} [1]
k_{Mg}	29.0 MPa/wt%^{3} [1]
$\Delta\sigma_{gb}$	16 MPa [1]
Taylor Factor M	3.1 [1]
r_c	5 nm [9]
G_{Al}	26.5 GPa [33]
G_{Mg2Si}	37.4 GPa [33]
$\Gamma = (G_{Al}b^2)/2$	1.1025×10^{-14} GPa × m^2 [33]

3. Results

3.1. Al-Mg Phase Diagrams

Figure 2 illustrates the binary Al-Mg phase diagrams, with constant Si concentration (wt.%) equal to 0.6 and 1.1. In both cases, for temperatures above 650 °C, the microstructure includes only the liquid phase while, during undercooling, the formation of iron-intermetallic phases has been observed at 620 °C. For the lower Si concentration (0.6 wt.%), the iron intermetallic phase has a general chemical formula of Al_8Fe_2Si, attributed to the general category of α-AlFeSi particles, having spherical morphology and cubic crystal structure. After the end of solidification and about the range 620–580 °C, Al_8Fe_{Si} will be transformed to $Al_9Fe_2Si_2$, attributed to the general category of β-AlFeSi particles, needle-shaped, having a hexagonal crystal structure [35,36]. On the contrary, for a higher Si concentration (1.1 wt.%) the transformation α-AlFeSi to β-AlFeSi is achieved before the completion of solidification. For each case, β-AlFeSi remains thermodynamically stable until room temperature. For the β-phase (Mg_2Si), the temperature range of formation depends on the Mg and Si concentration. For Mg concentration equal to 0.45%, Mg_2Si is thermodynamically stable from 450–500 °C, for Mg concentration equal to 0.9%, Mg_2Si is thermodynamically stable from 500–550 °C, while for Mg concentration equal to 1.0%, Mg_2Si is thermodynamically stable from 550–600 °C, just after the solidification.

The constructed phase diagrams confirmed the importance of Si content for the composition and temperature stability of iron-intermetallic phases. It is worth mentioning, that the microstructure phases, depicted in the constructed diagrams, are attributed to equilibrium conditions, far beyond those appearing during the conventional processing, as infinite time is given for the formation of phases like AlMn and diamond Si. Consequently, the aforementioned phases are not anticipated to participate during the ageing treatment. On the other side, the equilibrium conditions of phase diagrams reason the absence of metastable Mg_xSi_y phases. The quenching, after the end of annealing, prevents the formation of Mg_2Si phases at low temperatures, as described by the phase-diagrams. Consequently, it is imperative to conduct the ageing treatment above 150 °C for precipitates being able to nucleate, grow, and coarsen, but lower than 300 °C so as to prevent the recrystallisation of the matrix phase.

3.2. Recrystallisation and Grain Growth Simulation

The temporal evolution of recrystallized fraction is illustrated in Figure 3. It can be seen that the presence of pinning action affects recrystallisation kinetics, causing a slight deceleration of the mechanism as the fully recrystallized microstructure is achieved after t = 150 s, while the fully recrystallized microstructure is achieved without the action of nanoparticles after t = 100 s. Similarly, a slight deviation is noticed for the average radius of the recrystallized grains, Figure 4, where the action of pinning force causes an average reduction of about 1 μm. The difference between the two cases is noticed for the overall number of recrystallized grains. For both cases, the initial deformed microstructure consists of 230 grains. During the simulation, the number of grains is reduced, as the most developed grains move competitively through the interface of their neighbouring grains and grow at the expense of the smaller ones that shrink.

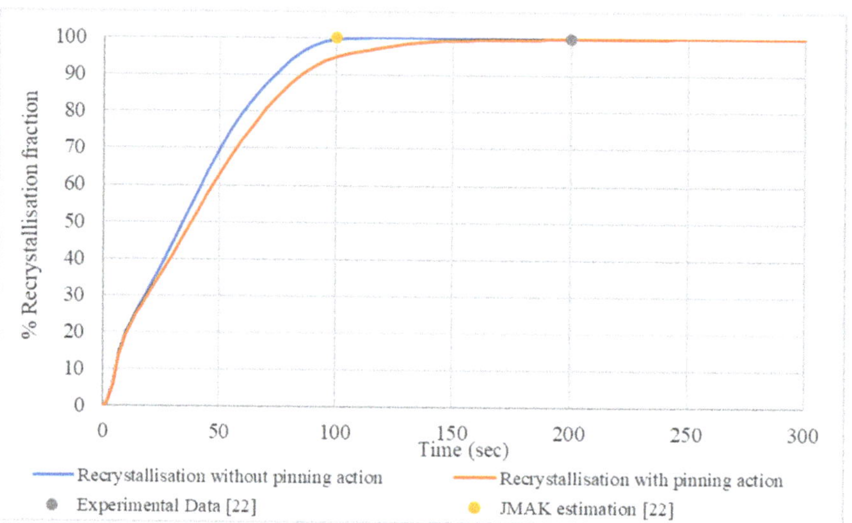

Figure 3. Evolution of recrystallisation fraction. The JMAK estimation and experimental data are taken from the work of Poletti et al. [22].

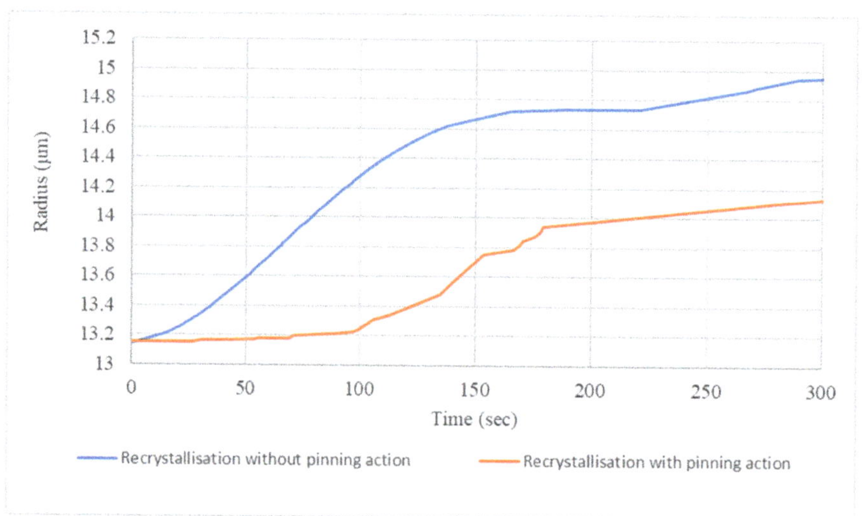

Figure 4. Evolution of the recrystallized grains average radius.

The presence of the secondary particles, through their pining impact, results in a lower decrease of the overall number of grains, the number of which is stabilized between 150–200 s. The evolution of the grain's number is depicted in Figure 5. The aforementioned differences between the two simulations can be explained by their difference in the average value of interface mobility, as illustrated in Figure 6. As anticipated, the pinning action causes lower values of interface mobility with the deviation of the two cases remaining constant throughout the simulation.

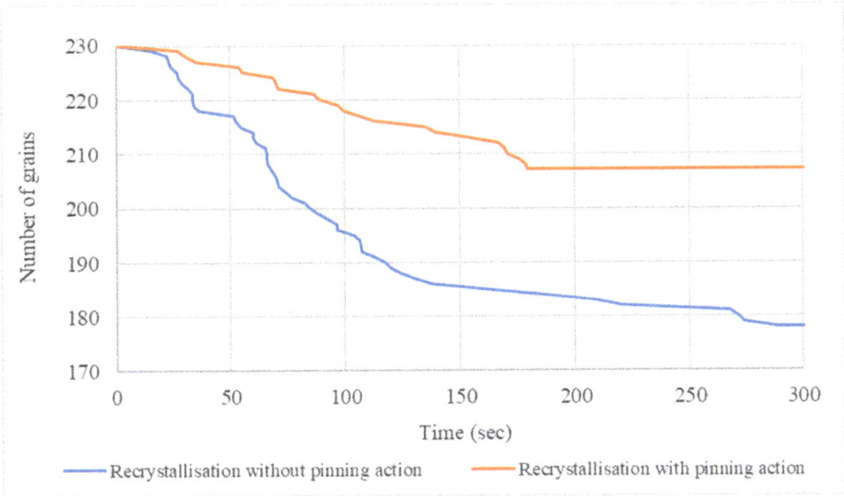

Figure 5. Evolution of the number of grains during the simulation.

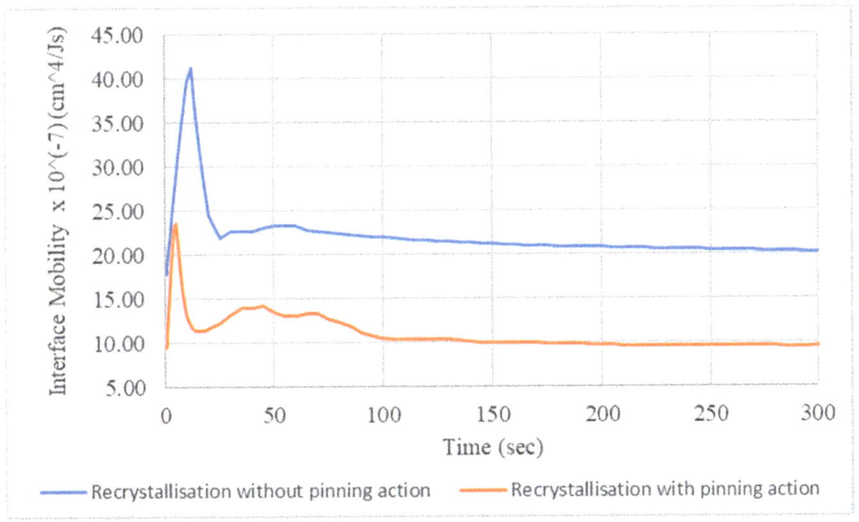

Figure 6. Evolution of interface mobility during the simulation.

3.3. Isothermal Artificial Ageing Simulation

Figure 7 depicts the mesh analysis for the 220 °C-4 h simulation of the Al-0.9Mg-0.6Si ternary system. For the lower values of cells (500–800–1100), severe deviation is estimated among the successive simulations, for the precipitate fraction, in the range between 20–30%. On the contrary, the analysis with 1700–2000 cells, appears to provide much more similar results, with a slight deviation, lower than 2%. Such a deviation can be thought of as acceptable, and consequently, between the two analyses, the 1700 × 1700 is selected due its lower overall simulation time. The analysis of 1700 × 1700 cells, with cell dimension equal to 11.8 nm and interface width equal to 35.4 nm (3 × 11.8), will be implemented for the subsequent ageing simulations. Choosing the 1700 × 1700 cells, Figure 8 presents the precipitate fraction for the 180–200 °C ageing simulations.

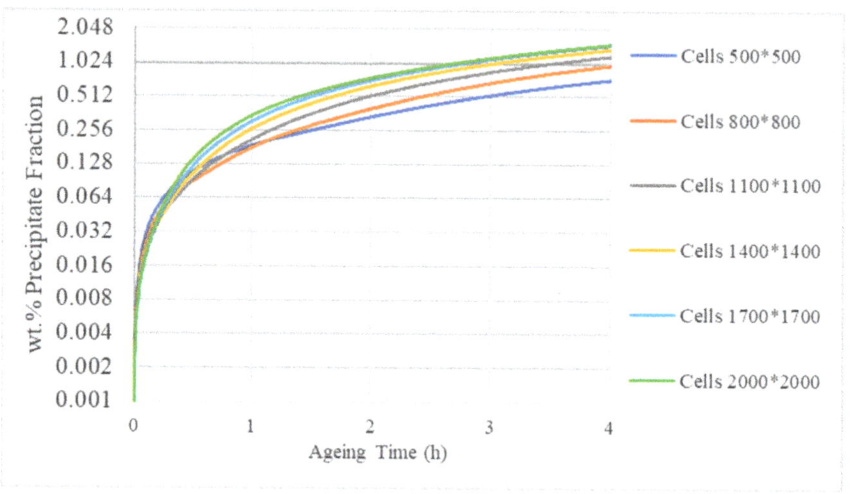

Figure 7. Mesh analysis of 220 °C-4 h ageing treatment for Al-0.9M-0.6Si ternary system.

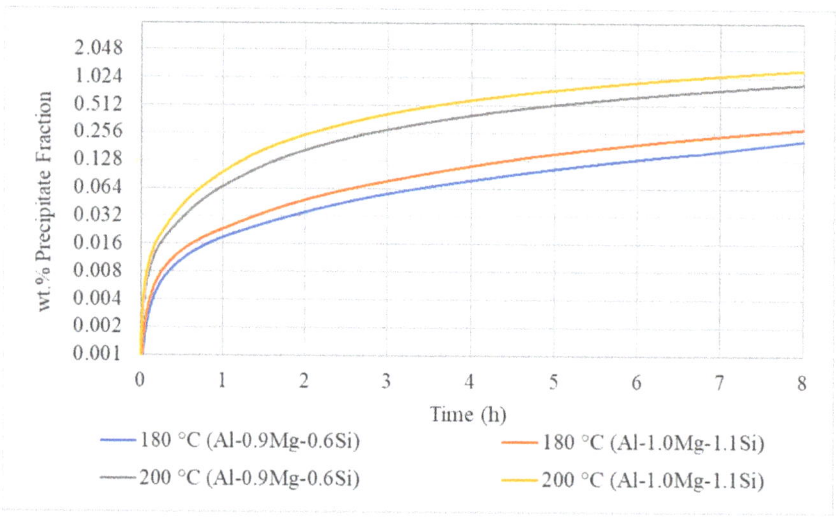

Figure 8. Temporal evolution of the precipitate fraction.

For the 180 °C ageing, the estimated wt.% of Mg_2Si is quite low, equal to 0.21 for the ternary Al-0.9Mg-0.6Si and 0.29 for Al-1.0Mg-1.1Si. For the higher ageing temperature of 200 °C, a significant increase is recorded as the precipitate fraction reaches the value 0.87 for the ternary Al-0.9Mg-0.6Si and the value 1.25 for the ternary system Al-1.0Mg-1.1Si.

As illustrated by Figure 9, the precipitate average radius records its lower values for the 180 °C ageing simulation, equal to 16.4 nm for Al-0.9Mg-0.6Si and 17 nm for Al-1.0Mg-1.1Si. For 200 °C ageing, the average radius significantly increases, reaching the value 32.2 nm for the ternary Al-0.9Mg-0.6Si and 37.9 nm for Al-1.0Mg-1.1Si system. For constant ageing temperature, the two compositions record similar values, with a slightly higher price achieved by the Al-1.0Mg-1.1Si system. From both Figures 9 and 10, it can be seen that for the 180 and 200 °C simulations, precipitates and their average radius are rapidly grown, while their increasing rate reaches a plateau after 4 h of ageing. Given that all simulations begin with the same number of Mg_2Si grains and the same overall study surface, Figure 10 depicts the evolution of precipitate's density.

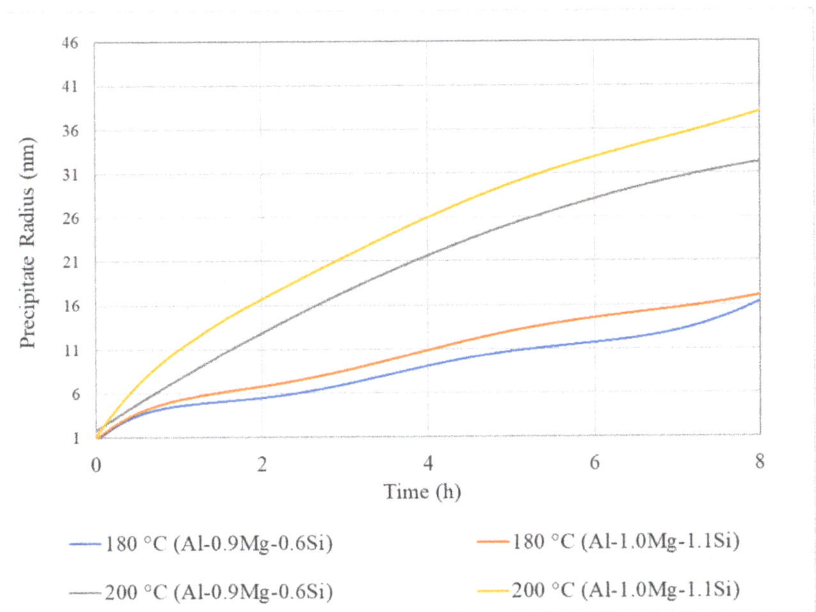

Figure 9. Temporal variation of the precipitate radius.

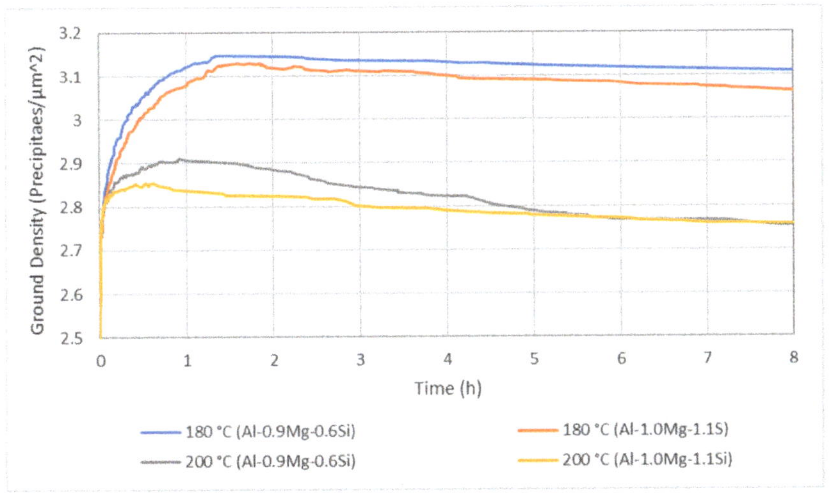

Figure 10. Temporal evolution of the precipitate's density.

It can be seen that, in every simulation, a slight decrease of precipitate density is recorded as the thermodynamically unstable precipitates are dissolved providing Mg and

Si content for the growth of the thermodynamically stable precipitates. It is also observed that for the lower ageing temperatures (180 °C), with the lowest precipitate fractions, precipitate density appears to have its highest values. On the other side, the 200 °C ageing simulations, attributed to the maximum precipitate fraction, have the lowest values of precipitate density. This contradiction between the precipitate fraction and density is explained by the fact that the rise of ageing temperature results in the increase of the average precipitate radius, although the overall number of precipitates is reduced.

Figure 11 depicts the variation of Mg and Si concentration in the matrix. Mg and Si concentrations in Mg$_2$Si are constant, 63% and 37%, respectively. The conducted simulations do not consider the iron intermetallic particles, which restrict the available Si, based on Equation (10), as long as the presence of iron intermetallic phases reduces the available Si for the Mg$_2$Si precipitation.

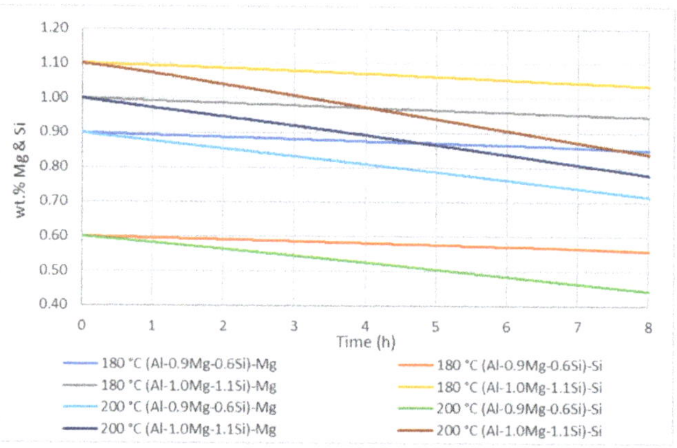

Figure 11. Temporal evolution of Mg and Si content in the matrix phase.

A posteriori, Equation (10) is implemented, taken the composition of Mn kai Fe from Table 1, so as to identify if the absence of iron particles resulted in an overestimation of available Si, and consequently, an overestimation of precipitate fraction and radius. Figure 12 presents the availability of Si for precipitation based on the restriction of Equation (10). As anticipated, Mg and Si concentrations in the matrix decreased, as the volume fraction of precipitates increased, with the most significant reduction estimated in the 200 °C ageing simulation, where the highest value of precipitates is recorded. The variation of Mg concentration in the matrix phase is most significant in comparison to Si, as Mg has higher concentration in Mg$_2$Si phase, rather than Si. Based on Figure 12, it can be estimated that the absence of iron phases in simulation microstructure does not lead to overestimation of precipitate volume fraction as Si concentration has a surplus in the matrix.

$$c_{eff}^{Si} = c_o^{Si} - 0.25\left(c_o^{Fe} + c_o^{Mn}\right) \tag{10}$$

Figure 13 presents the microstructure simulation by the end of each ageing simulation. It can be seen that there is a mutual distribution of precipitates, both on the bulk region of the matrix phase grains, the interfaces, and the triple junctions. At the initial ageing steps, particles precipitated primarily in the bulk region, while some precipitate free zones were present in the microstructure. Around the triple junctions, some precipitation-free zones have been noticed, as the formation of particles in the triple junctions depletes the availability of Mg and Si for further precipitation. Figure 14 presents the interface mobility, and Table 6 the diffusion coefficients of Mg and Si in matrix phase calculated by ThermoCalC$^@$. The interface mobility records its highest values for the lower temperature of 180 °C, and its lower value for the highest ageing temperature of 200 °C. On the contrary, diffusion coefficients of Mg and Si in the matrix phase are increased for higher ageing temperatures with the Si coefficient having a slightly higher value than its Mg counterpart. As the ageing temperature is increased, the diffusion mechanism is getting more important, while for lower ageing temperatures, the interface mechanism is of greater importance. Diffusion coefficients in precipitates are considered equal to zero, as Mg_2Si are stoichiometric intermetallic compounds with standard chemical composition and minimum solubility.

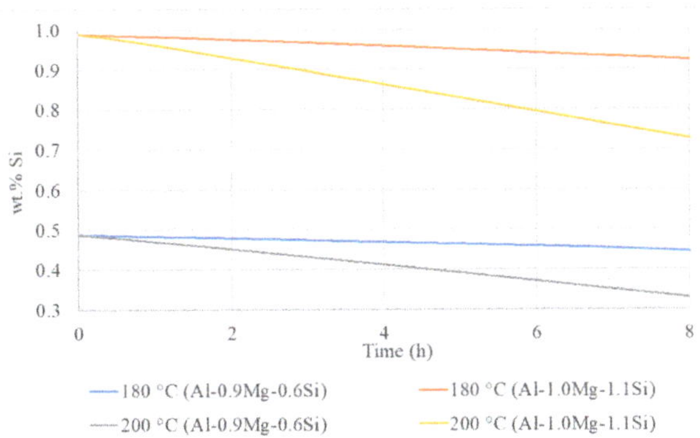

Figure 12. Evolution of available Si content in the matrix phase, based on Equation (10).

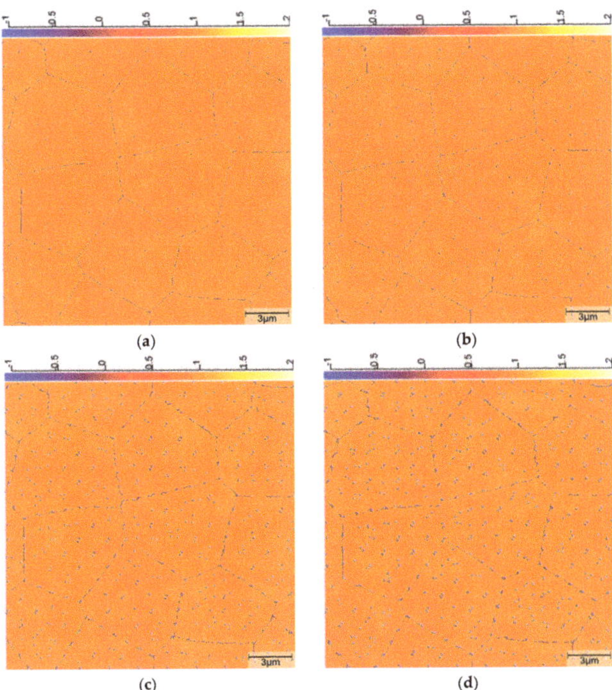

Figure 13. Spatial distribution of precipitates. The left column depicts the ternary system. Al-0.9Mg-0.6Si and the right column the Al-1.0Mg-1.1Si. Cases (**a**,**b**) are for the 180 °C-8 h simulation, while cases (**c**,**d**) for the 200 °C-8 h simulation. In the colour scale, number 1 depicts the matrix phase, number 2 depicts precipitates, while number 1 is attributed to the interfaces.

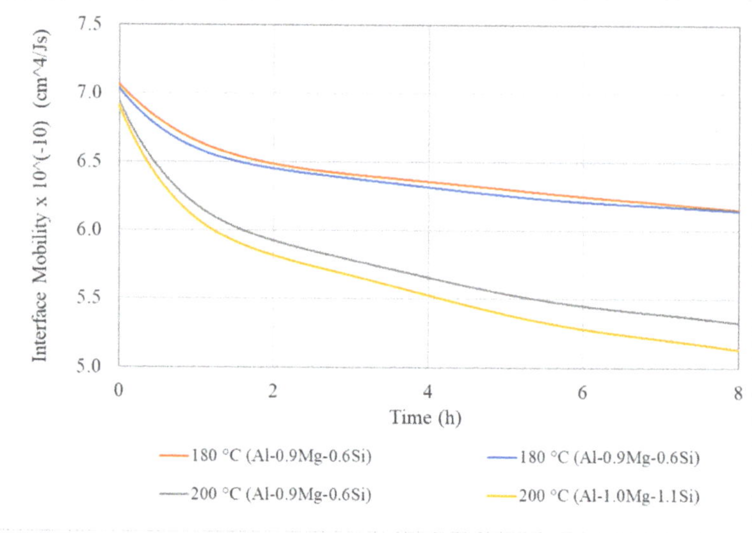

Figure 14. Temporal evolution of interface mobility.

Table 6. Diffusion coefficients of Mg and Si in the matrix phase.

Simulation	Mg Diffusion Coefficient (cm²/s)	Si Diffusion Coefficient (cm²/s)
180 °C (Al-0.9Mg-0.6Si)	2.40×10^{-15}	3.77×10^{-15}
180 °C (Al-1.0Mg-1.1Si)	2.35×10^{-15}	3.66×10^{-15}
200 °C (Al-0.9Mg-0.6Si)	9.08×10^{-15}	1.42×10^{-14}
200 °C (Al-1.0Mg-1.1Si)	9.14×10^{-15}	1.37×10^{-14}

Figure 15 depicts the virtual EDX analysis, provided by the MICRESS®software, where the red curve represents the Si concentration and the black curve represents the Mg concentration, in both matrix phase and precipitates. The virtual EDX analysis has been considered for the horizontal line on the centre of the microstructure. The vertical axis represents the % concentration of Mg and Si, while the horizontal axis represents the position in the microstructure. For the lower ageing temperature, of 180 °C, a sparse dispersion of both Mg and Si curves has been recorded, because of the low volume fraction of precipitates. It is worth mentioning that the peak of Si curve is equal to 37% and the peak of Mg curve is equal to 63%, both attributed to Mg_2Si precipitates. For the 200 °C simulations, intense distribution is recorded, due to the increase of precipitate fraction, while some peaks have been noticed, lower than 63% and 37%. These peaks are attributed to precipitate particles, which have not received their final chemical composition of Mg_2Si. Figures 16 and 17 depict the estimated values of yield strength and hardness, based on Equations (4)–(9), for the final ageing conditions. For the 180 °C ageing, lower values of yield strength and hardness are recorded, due to the relatively low volume fraction of precipitates. More precisely, for the ternary Al-0.9Mg-0.6Si system, the values of yield strength and hardness are equal to 139 MPa and 62 HV, respectively, while for the ternary Al-1.0Mg-1.1Si, the corresponding values are 176 MPa and 74 HV. The increase of the precipitate fraction results in the enhancement of yield strength and hardness for both ternary systems. At 200 °C, yield strength is 223 (Al-0.9Mg-0.6Si) and 291 (Al-1.0Mg-1.1Si) MPa, while hardness is 89 and 112 HV, respectively.

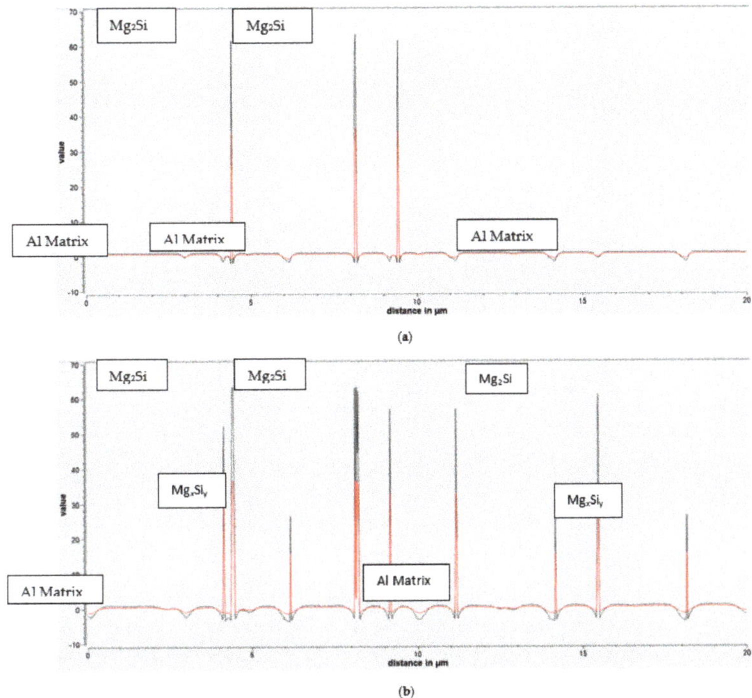

Figure 15. Virtual EDX analysis for the ternary Al-0.9Mg-0.6Si system for: 180 (**a**) and 200 °C (**b**). The horizontal axis depicts the position in the central line of the microstructure, while the vertical axis depicts the corresponding wt.% composition in Mg (black curve) and Si (red curve). The white boxes represent the corresponding phase for each couple of Mg-Si concentrations.

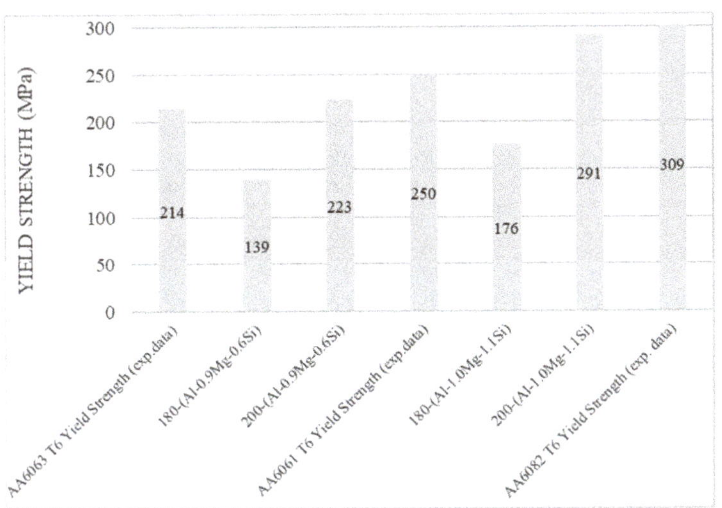

Figure 16. Yield strength estimation. Experimental data (AA6061-AA6063-AA6082) from [1,4,37].

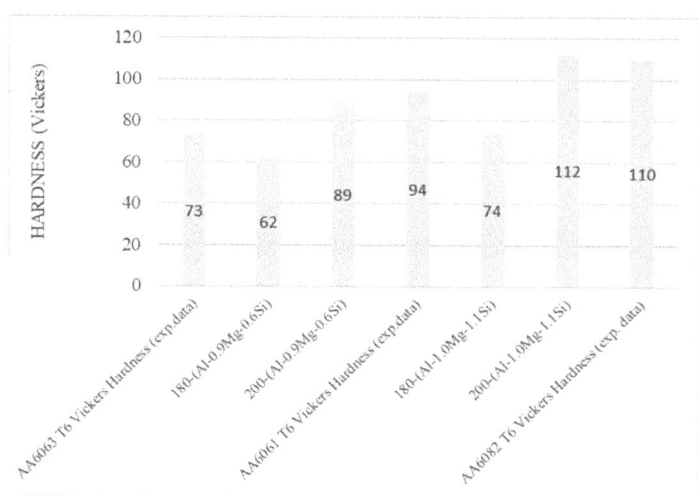

Figure 17. Hardness estimation. Experimental data (AA6061-AA6063-AA6082) from [4,9,37].

4. Discussion

The predicted kinetics of recrystallisation, by the phase-filed simulation, satisfactorily approaches the experimental data of Poletti et al. [22] for the recrystallisation of AA6XXX aluminium alloys, with the same level of previous deformation and similar impact of secondary particles. The insignificant impact of secondary particles can be explained by their relatively small volume fraction and the fact that only a small proportion of these particles is nanoscopic, retarding the recrystallisation mechanism, while the larger particles, in the range of some µm, enhance recrystallisation kinetics, through precipitation stimulated nucleation (PSN).

Regarding the ageing simulation, the particularly high number of cells in the microstructure analysis proves the difficulty in accurately illustrating nanoscopic particles as part of the overall microstructure. For the 180 °C ageing simulation, the 8h treatment does not secure the maximum expected value of precipitate fraction. However, there is a severe deviation, in literature, about the time required for the peak ageing conditions in 180 °C. Although the majority of works report an ageing time between 6–10 h [33,38], some reports refer to prolong ageing treatments [11], while ASTM [38] predicted 70 h of treatment for lower ageing temperatures, 170–180 °C. On the other side, the 200 °C simulation was in good agreement with the expected duration of the treatment. The variation of ageing kinetics between the 180 °C and 200 °C simulations can be explained by the increasement of diffusion rates of Mg and Si in 200 °C, which resulted in the increase of growth rate for the precipitates. The maximum values of the precipitate fraction agree, generally, with the ones expected by the literature. For the ternary Al-0.9Mg-0.6Si, which is within the Mg and Si concentration range of AA6061 and AA6063 alloys, the precipitate fraction lies on the range of 0.7 to 1.4% [1,4]. On the other side, the ternary Al-1.0Mg-1.1Si system is attributed to the AA6082 aluminium alloy, where the precipitate fraction can reach higher peaks, even nearby 1.4 to 1.9% [1,39]. The increased diffusion coefficients of 200 °C explain the significant increase of the average precipitate radius, as depicted in Figure 9. The gradual reduction of the overall number of precipitates is explained by the Ostwald ripening mechanism. Precipitates, whose radius is close to interface width, are unstable for the saturated matrix phase, where the elements tend to leave the lattice, so as to create more stable particles.

It is worth noting that the predicted values of precipitate radius appear to have a significant variation than those recorded for the Kampman–Wagner simulations, where the average precipitate radius is about 3 to 10 nm [1,9].

This deviation can be explained by the different approach toward the nucleation-growth-coarsening/dissolution of precipitates, the different values of critical ageing parameters, like the interface energy and the molar volume of the matrix phase and precipitates, the different approach toward the concentration gradient at the matrix phase-precipitates interfaces, and the fact that the applied Phase-Field model inserts atomic and mobility databases through CALPHAD modelling (ThermoCalC$^{@}$). It is worth mentioning that the final average radius of precipitates in the applied Phase-Field modelling is affected by the competitive relationship between particles, created during different ageing time steps [40–42]. The deviation of the average radius, between models, may result from the difference in the estimated values of activation energy for Mg and Si diffusion rates, as Bahrami [1] has shown that the reduction of the considered activation energy results in the increment of mean radius, due to the rise of the diffusion rates. According to literature, the average size of the metastable particles lies between 40–130 nm, depending on the morphology of precipitates and ageing conditions (temperature and time). Therefore, the simulated results are within the expected spectra [43,44].

The predicted values of hardness and yield strength for the 180 °C are far lower than those anticipated for the peak-ageing condition (T6) and agree with the under-ageing condition for the ternary Al-0.9Mg-0.6 and Al-1.0Mg-1.1Si system, while the 200 °C values approach the peak ageing condition. More precisely, the peak-ageing strength is about 214 (AA6063)-250(AA6061) [1,2,4] MPa for the ternary Al-0.9Mg-0.6Si system and 300–310 (AA6082) MPa [37,41,45] for the ternary Al-1.0Mg-1.1Si system. The peak ageing value of hardness is equal to 73–94 HV [2,4,9] for the ternary Al-0.9Mg-0.6Si, and 100–110 HV [37,41] for the ternary Al-1.0Mg-1.1Si system. The root causes of the lower values of hardness and strength in the 180 °C ageing treatment can be attributed to the lower diffusion rates in the specific temperature resulting in lower chemical driving force for the precipitation. It is worth noting that in the yield strength and hardness model, a rather conservative value for the annealing-intrinsic yield strength (16 MPa) has been used so as to primarily emphasize the importance of precipitation procedure for the enhancement of the mechanical properties. The value of the intrinsic strength combined with the absence of alloying elements contribution to the solid solution strengthening, beyond the contribution of Mg and Si, may result in a slight underestimation of the peak ageing yield strength and hardness.

Finally, the estimation for the uniform distribution of precipitates both in the bulk region of matrix phase, the interfaces and the triple junctions agree with the theory of heterogeneous nucleation and it is desired for the final microstructure, as the exclusive precipitate in the interfaces would made them intensively brittle, while the presence of Precipitate-Free-Zones (PFZ) [46] is connected with the lower corrosion resistance of the aged alloy, the rise of the chemical heterogeneity and the intergranular cracking, due to the variation of strength between the interior of the grains, which are enriched in particles, and the depleted interfaces.

5. Conclusions

The simulation of Al-Mg-Si alloys heat treatment through MultiPhase-Field based MICRESS$^{@}$ software, led to the following conclusions:

1. The 400 °C/5 min annealing simulation accurately predicted the recrystallisation kinetics proving a slight impact of secondary nanoparticles on the deceleration of recrystallisation mechanism and the average radius of recrystallized grains. This deceleration is explained by the lower values of the interface mobility.
2. The ageing simulation predicted the under-ageing condition for the 180 °C-8 h treatment and the peak ageing condition for the 200 °C-8 h ageing simulations.
3. For lower ageing temperatures, the interface mobility has more significant impact on the precipitation mechanism. On the contrary, the rise of temperature results in severe increase of diffusion mechanism, causing the coarsening of precipitate particles, which

nucleate and grow both on the interior of Al matrix phase grains and the interfaces and triple junctions.

Author Contributions: We describe contributions to the paper using the CRediT (Contributor Roles Taxonomy). Conceptualization: S.P.; methodology: S.P., A.B., and M.B.; software: A.B., and M.B.; validation: A.B. and M.B.; formal analysis: A.B. and M.B.; investigation: A.B. and M.B.; resources: A.B. and M.B.; data curation: A.B., M.B.; writing—original draft preparation: A.B.; writing—review and editing: A.B., M.B., and S.P.; visualization: A.B., M.B.; supervision: M.B. and S.P.; project administration: S.P. All authors have read and agreed to the published version of the manuscript.

Funding: This research did not receive any specific grant from funding agencies in the public, commercial, or non-for-profit sectors.

Institutional Review Board Statement: Not Applicable.

Informed Consent Statement: Not Applicable.

Data Availability Statement: Data available on request due to restrictions, as the research is ongoing.

Acknowledgments: The work described in this article was supported by the Hellenic Research Centre for Metals S.A (ELKEME S.A). The authors would like to acknowledge E. Gavalas for his assistance and scientific advice during the designing of the project.

Conflicts of Interest: The authors declare no conflict of interest.

References

1. Bahrami, A. Modeling of Precipitation Sequence and Ageing Kinetics in Al-Mg-Si Alloys. Ph.D. Thesis, Delft University of Technology, Delft, Poland, 2010.
2. Mukhopadhyay, P. Alloy Designation, Processing, and Use of AA6XXX Series Aluminium Alloys. *ISRN Metall.* **2012**. [CrossRef]
3. Engler, O.; Hirsch, J. Texture control by thermomechanical processing of AA6xxx Al–Mg–Si sheet alloys for automotive applications—A review. *Mater. Sci. Eng. A* **2002**, *336*, 249–262. [CrossRef]
4. Nisaratanaporn, E. Microstructural Development and Pressure Requirements in 6063 Aluminium Alloy Tube Extrusion. Ph.D. Thesis, Imperial College, London, UK, 1995.
5. Esmaeili, S.; LIoyd, D.J.; Poole, W.J. Modeling of precipitation hardening for the naturally aged Al-Mg-Si-Cu alloy AA6111. *Acta Mater.* **2003**, *51*, 3467–3481. [CrossRef]
6. Lan, Y.; Pinna, C. Modelling the static recrystallisation texture of FCC metals using a phase field method. *Mater. Sci. Forum* **2012**, *715*, 739–744. [CrossRef]
7. Jou, H.-J.; Lusk, M.T. Comparison of Johnson-Mehl-Avrami-Kologoromov kinetics with a phase-field model for microstructural evolution driven by substructure energy. *Phys. Rev. B* **1997**, *55*, 8114–8121. [CrossRef]
8. Weinkamer, R.; Fratzl, P.; Gupta, H.S.; Penrose, O.; Lebowitz, J.L. Using Kinetic Monte Carlo simulations to study phase separation in Alloys. *Phase Transit.* **2004**, *77*, 433–456. [CrossRef]
9. Myhr, O.R.; Grong, Ø.; Andersen, S.J. Modelling of the age hardening behaviour of Al–Mg–Si alloys. *Acta Mater.* **2001**, *49*, 65–75. [CrossRef]
10. Mao, F.; Bollmann, C.; Brüggemann, T.; Liang, Z.; Jiang, H.; Mohles, V. Modelling of the Age-Hardening Behavior in AA6xxx within a Through-Process Modelling Framework. *Mater. Sci. Forum* **2017**, *877*, 640–646. [CrossRef]
11. Povoden-Karadeniz, E.; Lang, P.; Warczok, P.; Falahati, A.; Jun, W.; Kozeschnik, E. Calphad modeling of metastable phases in the Al–Mg–Si system. *Calphad* **2013**, *43*, 94–104. [CrossRef]
12. Raabe, D. *Computational Materials Science: The Simulation of Materials Microstructures and Properties*; Wiley-VCH: New York, NY, USA, 1998.
13. Moelans, N.; Blanpain, B.; Wollants, P. An introduction to phase-field modeling of microstructure evolution. *Calphad* **2007**, *32*, 268–294. [CrossRef]
14. Bhadeshia, H.; Qin, R.S. Phase field method. *Mater. Sci. Technol.* **2010**, *26*, 803–811.
15. Lewis, D.; Warren, J.; Boettinger, W.; Pusztai, T.; Granasy, L. Phase-field models for eutectic solidification. *Jom* **2004**, *56*, 34–39. [CrossRef]
16. Nestler, B.; Choudhury, A. Phase-field modeling of multi-component systems. *Curr. Opin. Solid State Mater. Sci.* **2011**, *15*, 93–105. [CrossRef]
17. Mueller, J.J.; Matlock, D.K.; Speer, J.; De Moor, E.; Moor, D. Accelerated Ferrite-to-Austenite Transformation During Intercritical Annealing of Medium-Manganese Steels Due to Cold-Rolling. *Metals* **2019**, *9*, 926. [CrossRef]
18. Ta, N.; Zhang, L.; Du, Y. Design of the Precipitation Process for Ni-Al Alloys with Optimal Mechanical Properties: A Phase-Field Study. *Metall. Mater. Trans. A* **2014**, *45*, 1787–1802. [CrossRef]

19. Xu, Q.; Zhang, Y. Precipitation and Growth Simulation of γ' Phase in Single Crystal Superalloy DD6 with Multiphase-Field Method and Explicit Nucleation Algorithm. *Metals* **2020**, *10*, 1346. [CrossRef]

20. Ji, Y.; Ghaffari, B.; Li, M.; Chen, L.-Q. Phase-field modeling of θ' precipitation kinetics in 319 aluminum alloys. *Comput. Mater. Sci.* **2018**, *151*, 84–94. [CrossRef]

21. Steinbach, I.; Pezzolla, F.; Nestler, B.; Seeßelberg, M.; Prieler, R.; Schmitz, G.J.; Rezende, J.L. A phase field concept for multiphase systems. *Phys. D Nonlinear Phenom.* **1996**, *94*, 135–147. [CrossRef]

22. Poletti, M.C.; Bureau, R.; Loidolt, P.; Simon, P.; Mitsche, S.; Spuller, M. Microstructure Evolution in a 6082 Aluminium Alloy during Thermomechanical Treatment. *Material* **2018**, *11*, 1319. [CrossRef]

23. Pedersen, K.O.; Lademo, O.-G.; Berstad, T.; Furu, T.; Hopperstad, O. Influence of texture and grain structure on strain localisation and formability for AlMgSi alloys. *J. Mater. Process. Technol.* **2008**, *200*, 77–93. [CrossRef]

24. Georgakou, A.-G. Phase-Field Simulation of Recrystallization and Grain Growth: Description of MICRESS Software and Application in AI-Mg-Sc-Zr Alloy. Bachelor's Thesis, University of Volos, Volos, Greece, 2013.

25. Ratchev, P.; Jessner, P. Role of the Mn-Dispersoids and Mg_2Si Particles in the Recrystallization of Automotive 6xxx Alloys. *Mater. Sci. Forum* **2014**, *794*, 1227–1232. [CrossRef]

26. Farè, S.; Lecis, N.; Vedani, M. Aging Behaviour of Al-Mg-Si Alloys Subjected to Severe Plastic Deformation by ECAP and Cold Asymmetric Rolling. *J. Met.* **2011**, *2011*, 1–8. [CrossRef]

27. Adamczyk-Cieślak, B.; Mizera, J. Ultra-Fine Grain Structures of Model Al-Mg-Si Alloys Produced by Hydrostatic Extrusion. In Proceedings of the International Conference on Advances in Materials and Processing Technologies, Paris, France, 24–27 October 2010.

28. Kai, X.; Chen, C.; Sun, X.; Wang, C.; Zhao, Y. Hot deformation behavior and optimization of processing parameters of a typical high-strength Al–Mg–Si alloy. *Mater. Des.* **2016**, *90*, 1151–1158. [CrossRef]

29. Werinos, M.; Antrekowitsch, H.; Kozeschnik, E.; Ebner, T.; Moszner, F.; Löffler, J.; Uggowitzer, P.; Pogatscher, S. Ultrafast artificial aging of Al–Mg–Si alloys. *Scr. Mater.* **2016**, *112*, 148–151. [CrossRef]

30. Yang, M.; Wei, H.; Zhang, J.; Zhao, Y.; Jin, T.; Liu, L.; Young, S. Phase-field study on effects of antiphase domain and elastic energy on evolution of γ' precipitates in nickel-based superalloys. *Comput. Mater. Sci.* **2017**, *129*, 211–219. [CrossRef]

31. Fallah, V.; Korinek, A.; Raeisinia, B.; Gallerneault, M.; Esmaeili, S. Early-Stage Precipitation Phenomena and Composi-tion-dependent Hardening in Al-Mg-Si-(Cu) Alloys. *Mater. Sci. Forum* **2014**, *794–796*, 933–938. [CrossRef]

32. Qiao, J.; Chen, J.; Che, H. Crashworthiness assessment of square aluminum extrusions considering the damage evolution. *Thin-Walled Struct.* **2006**, *44*, 692–700. [CrossRef]

33. Yang, M.; Chen, H.; Orekhov, A.; Lu, Q.; Lan, X.; Li, K.; Zhang, S.; Song, M.; Kong, Y.; Schryvers, D.; et al. Quantified contri-bution of β'' and β' precipitates to the strengthening of an aged Al–Mg–Si alloy. *Mater. Sci. Eng. A* **2020**, *774*, 138776. [CrossRef]

34. Nandy, S.; Ray, K.K.; Das, D. Process model to predict yield strength of AA6063 alloy. *Mater. Sci. Eng. A* **2015**, *644*, 413–424. [CrossRef]

35. Sarafoglou, P.I.; Serafeim, A.; Fanikos, I.A.; Aristeidakis, J.S.; Haidemenopoulos, G.N. Modeling of Microsegregation and Homogenization of 6xxx Al-Alloys Including Precipitation and Strengthening During Homogenization Cooling. *Material* **2019**, *12*, 1421. [CrossRef]

36. Kuijpers, N.C.; Vermolen, F.J.; Vuik, C.; Van der Zwang, S. A Model of the β-AlFeSi to α-Al(FeMn)Si transformation in Al-Mg-Si Alloys. *Mater. Trans.* **2003**, *44*, 1448–1456. [CrossRef]

37. Liang, Y. An Investigation on the Optimum Aging Condition for HFQ-processed AA6082 Aluminium Alloy. Master's Thesis, University of Birmingham, Birmingham, UK, 2016.

38. Edwards, G.A.; Stiller, K.; Dunlop, G.L.; Couper, M.J. The precipitation sequence in Al-Mg-Si alloys. *Acta Mater.* **1998**, *46*, 3893–3904. [CrossRef]

39. Herrnring, J.; Kashaev, N.; Klusemann, B. Precipitation Kinetics of AA6082: An Experimental and Numerical Investigation. *Mater. Sci. Forum* **2018**, *941*, 1411–1417. [CrossRef]

40. Wagner, R.; Kampman, R. *Materials Science and Technology-A Comprehensive Treatment*; Wiley-VCH: Weinheim, Germany, 1991; p. 5.

41. Du, Q.; Tang, K.; Marioara, C.D.; Andersen, S.J.; Holmedal, B.; Holmestad, R. Modeling over-ageing in Al-Mg-Si alloys by a multi-phase Calphad-coupled Kampmann-Wagner Numerical model. *Acta Mater.* **2017**, *122*, 178–186. [CrossRef]

42. Holmedal, B.; Osmundsen, E.; Du, Q. Precipitation of Non-Spherical Particles in Aluminum Alloys Part I: Generalization of the Kampmann–Wagner Numerical Model. *Met. Mater. Trans. A* **2016**, *47*, 581–588. [CrossRef]

43. Parson, N.C.; Yiu, H.L. The effect of heat treatment on the microstructure and properties of 6000 series alloy extrusion ingots. *Miner. Met. Mater. Soc.* **1989**, *40*, 18.

44. Camero, S.; Puchi, E.S.; Gonzalez, G. Effect of 0.1% vanadium on precipitation behavior and mechanical properties of Al-6063 commercial alloy. *J. Mater. Sci.* **2006**, *41*, 7361–7373. [CrossRef]

45. Polmer, I. *Light Alloys: From Traditional Alloys to Nanocrystals*, 4th ed.; Elsevier: Amsterdam, The Netherlands, 2005.

46. Eastering, K.; Porter, D. *Phase Transformations in Metals*; Chapman & Hall: Hong Kong, China, 1992.

Article

Role of Non-Metallic Inclusions in the Fracture Behavior of Cold Drawn Pearlitic Steel

Jesús Toribio *, Francisco-Javier Ayaso and Beatriz González

Fracture & Structural Integrity Research Group (FSIRG), Campus Viriato, University of Salamanca (USAL), EPS, Avda. Requejo 33, 49022 Zamora, Spain; fja@usal.es (F.-J.A.); bgonzalez@usal.es (B.G.)
* Correspondence: toribio@usal.es; Tel.: +34-677566723

Abstract: In this paper an exhaustive scientific work is performed, by means of metallographic and scanning electron microscope (SEM) techniques, of the microstructural defects exhibited by pearlitic steels and their evolution with the manufacturing process by cold drawing, analyzing the consequences of such defects on the *isotropic/anisotropic fracture behavior* of the different steels. Thus, the objective is the establishment of a relation between the microstructural damage and the fracture behavior of the different steels. To this end, samples were taken from all the intermediate stages of the real cold drawing process, from the initial hot rolled bar (not cold drawn at all) to the heavily drawn final commercial product (prestressing steel wire). Results show the very relevant role of non-metallic inclusions in the fracture behavior of cold drawn pearlitic steels.

Keywords: pearlitic steel; cold drawing; second-phase particles; non-metallic inclusions

Citation: Toribio, J.; Ayaso, F.-J.; González, B. Role of Non-Metallic Inclusions in the Fracture Behavior of Cold Drawn Pearlitic Steel. *Metals* **2021**, *11*, 962. https://doi.org/10.3390/met11060962

Academic Editor: Spyros Papaefthymiou

Received: 30 April 2021
Accepted: 11 June 2021
Published: 15 June 2021

Publisher's Note: MDPI stays neutral with regard to jurisdictional claims in published maps and institutional affiliations.

1. Introduction

High-strength cold-drawn eutectoid steel wires are important components in structural engineering [1–6]. As a consequence of the manufacture process by cold drawing, these materials show an anisotropic behavior with regard to plasticity (yielding), fracture and hydrogen embrittlement [7–18] with the result of mixed-mode fracture propagation and *strength anisotropy*.

The microstructural evolution with cold drawing has been extensively studied in the past [19–27], showing progressive microstructural orientation (alignment in a direction quasi-parallel to the wire axis or cold drawing direction) and increase of packing closeness associated with a decrease of interlamellar spacing and an orientation of the plates in the cold drawing direction [22–25].

A materials-science relationship between microstructure and strength is usually established through the Hall-Petch equation [28–31] to correlate the pearlite interlamellar spacing and the material strength. However, in oriented pearlitic microstructure (as a consequence of manufacturing by cold drawing) the Hall–Petch equation does not perform very well [32], so that an Embury–Fisher equation has been proposed [33] to describe the role of microstructure in material performance.

The role of inclusions in steel performance has received considerable attention in the scientific community, from studies about metallographic techniques [34] to research about models of void growth around inclusions [35,36], models about the spatial distribution of MnS inclusions and the voids provoked by them [37], or determining the volume fraction of inclusions in steel [38], as well as research on the role of inclusions in fatigue behavior in air [39,40] and hydrogen environment [41,42].

Recent references about inclusions in steel deal with modeling inclusion formation [43], effect of sulfur content [44], the effect of different non-metallic inclusions on the machinability of steels [45], quality control of steel wires [46], micromechanical modeling of fatigue crack nucleation around non-metallic inclusions [47], MnS inclusions formation in resulfu-

69

rised steel [48], influence of inclusions on the mechanical properties [49] or their effect on deformation and fracture [50].

This paper focuses on the role of inclusions (analyzed by metallographic techniques) in the mechanical performance (evaluated by means of standard tension tests) of cold drawn pearlitic steel with different degree of cold drawing and distinct chemical composition. The aim of the paper is to find a possible relationship (in the sense of materials science and engineering) between the microstructural micro-damage in the pearlitic steel (created by the presence of inclusions) and the fracture behavior of cold drawn pearlitic steels during tensile testing.

2. Materials

The samples used in the mechanical tests were eutectoid steel wires with different level of cold drawing (i.e., distinct degrees of accumulated plastic strain), from the initial hot rolled bar (not cold drawn at all) to the final commercial product (prestressing steel wire; heavily cold drawn), all of them corresponding to real manufacturing chains, as shown in Figures 1 and 2 that include different views of the process.

Figure 1. Manufacture of prestressing steel wires by progressive cold drawing of a previously hot-rolled pearlitic steel bar: two views of a real manufacturing chain by progressive (multi-step) cold drawing.

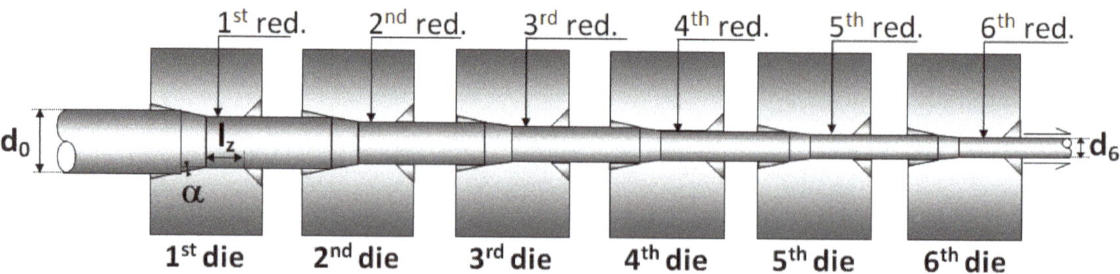

Figure 2. Schematic presentation of the cold drawing process with multiple passes through the dies.

Table 1 shows the chemical composition (wt.%) of the different steels (valid for any drawing degree). The materials used were cold drawn pearlitic steels corresponding to five real manufacturing processes. The five types (or families) of steels were used, with the names A, B, C, D and E, each of them with a specific chemical composition and manufacturing route (straining path). Steel of the type A undergoes six drawing steps (as shown in Figure 2 passing through six dies), whereas the others (B, C, D, E) have undergone seven drawing steps. A letter followed by a number will be used throughout this paper to indicate the steel family (the letter) and the degree of cold drawing or number of drawing steps undergone by each steel wire in particular.

Table 1. Chemical composition (wt.%) for the five families of the steels (the balance is Fe).

Steel	Family A	Family B	Family C	Family D	Family E
% C	0.80	0.789	0.79	0.795	0.789
% Mn	0.69	0.698	0.670	0.624	0.681
% Si	0.23	0.226	0.20	0.224	0.21
% P	0.012	0.011	0.009	0.011	0.010
% S	0.009	0.005	0.009	0.008	0.008
% Al	0.004	0.003	0.003	0.003	0.003
% Cr	0.265	0.271	0.187	0.164	0.218
% V	0.06	0.078	0.053	0.064	0.061

Table 2 summarizes each one of the five cold drawing procedures (*straining paths* or *yielding histories*) in terms of the wire diameter at the end of each drawing step. Table 3 offers the cumulative plastic strain ε^P_{cum} as the variable representing the drawing degree, and it's defined as follows [4]:

$$\varepsilon^P_{cum} = 2 \ln \frac{\phi_0}{\phi_i} \tag{1}$$

where ϕ_0 is the hot rolled bar diameter and ϕ_i is the diameter of a wire undergoing i dra-wing steps.

Table 2. Diameter of the wires at the end of each drawing step for the five families.

Drawing Step	Wire Diameter (mm)				
	Family A	Family B	Family C	Family D	Family E
0	12.11	12.10	10.44	8.56	11.03
1	10.80	11.23	9.52	7.78	9.90
2	9.81	10.45	8.49	6.82	8.95
3	8.94	9.68	7.68	6.17	8.21
4	8.22	9.02	6.95	5.61	7.49
5	7.56	8.54	6.36	5.08	6.80
6	6.98	8.18	5.86	4.63	6.26
7	-	7.00	5.03	3.97	5.04

Table 3. Cumulative plastic strain of the progressively drawn steels for the five families [4].

Drawing Step	ε^P_{cum}				
	Family A	Family B	Family C	Family D	Family E
0	0	0	0	0	0
1	0.229	0.149	0.184	0.191	0.216
2	0.421	0.293	0.414	0.454	0.418
3	0.607	0.446	0.614	0.655	0.591
4	0.775	0.588	0.814	0.845	0.774
5	0.942	0.697	0.991	1.044	0.967
6	1.102	0.800	1.155	1.229	1.133
7	-	1.095	1.460	1.537	1.566

The mechanical response of the cold drawn pearlitic steel wires is associated with a progressive increase with cold drawing of the yield strength (σ_Y) and of the ultimate tensile strength (σ_R), as shown in Figures 3 and 4.

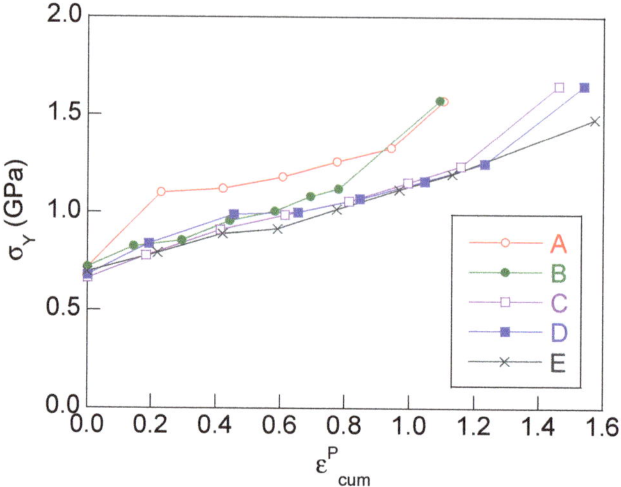

Figure 3. Evolution with cold drawing of the yield strength σ_Y.

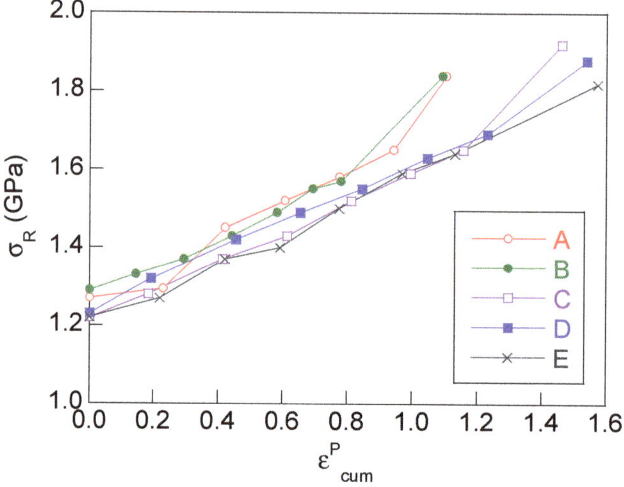

Figure 4. Evolution with cold drawing of the ultimate tensile strength σ_R.

Figure 5 plots the stress–strain curves for the progressively cold-drawn pearlitic steels from A0 (hot rolled steel, not cold drawn at all, 0 drawing steps) to the commercial prestressing steel wire A6 (heavily cold drawn pearlitic steel that has undergone 6 drawing steps). It is seen that both the yield strength σ_Y and the ultimate tensile strength (UTS σ_R) increase with the drawing degree [1–6].

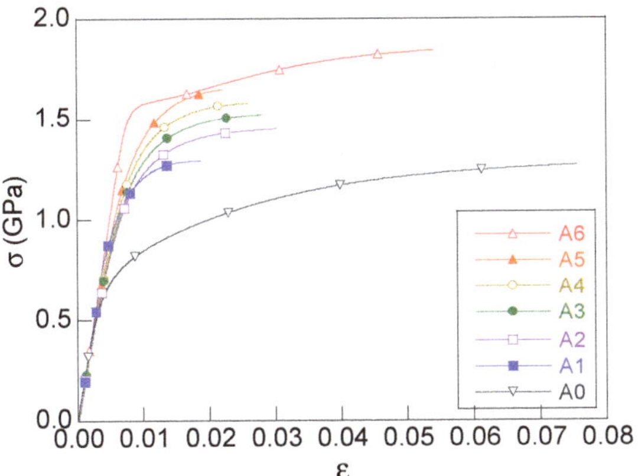

Figure 5. Stress–strain curves of the progressively drawn pearlitic steels A0 to A6 (steel family A: from 0 to 6 cold-drawing steps).

3. Metallographic Analysis

3.1. Sample Preparation

In order to proceed with the metallographic analysis of the different steels, representative samples were extracted from the wires by means of transverse cuts to prepare small cylinders of 10 mm height and the exact diameter of each steel wire in particular. After this, the cylinders underwent a cut in longitudinal direction (axial cut), so that the micrographs of the present paper will always be oriented with their vertical side following the cold drawing direction (wire axis).

The samples were mounted in resin for grinding and polishing up to mirror quality. Finally, they were chemically attacked to reveal the microstructure of the materials (progressively cold drawn pearlitic steel wires) and to observe it by scanning electron microscopy (SEM) using a JEOL JSM-5610 LV (Jeol Ltd., Tokyo, Japan). Etching was produced by using a solution of 4% Nital in chemical ethanol for five seconds.

3.2. Metallographic Observation

The chemical attack (etching) on the polished samples of pearlitic microstructure produces different chemical reactions in the *ferrite* (softer) phase and in the *cementite* (harder) phase forming the *pearlite*. Whereas the former is chemically attacked by the Nital, the latter remains unaffected by it, so that it is possible to distinguish both phases of the pearlitic microstructure of the steels by means a scanning electron microscope (SEM) analysis: cementite lamellae exhibit a clear appearance whereas ferrite lamellae show a darker aspect. The inclusions were observed by using energy-dispersive X-ray spectroscopy (EDX, Oxford Instruments, mod. 6587, High Wycombe, England) assembled to the SEM.

Generally speaking (details will be analyzed in the discussion section of the present paper), second-phase particles (*inclusions*) appear in *all* the studied pearlitic steels (five families or groups), exhibiting different peculiarities depending on the particular chemical composition of each family, cf. Table 1. Different types of inclusion were found, namely: (i) manganese sulfur (MnS) inclusions with dark (non-brilliant) appearance and irregular shape; (ii) silica (SiO_2) and aluminum oxides (Al_2O_3) with clear brilliant appearance and more regular shape. Two of these types of inclusions (MnS and Al_2O_3) are showed in Figure 6.

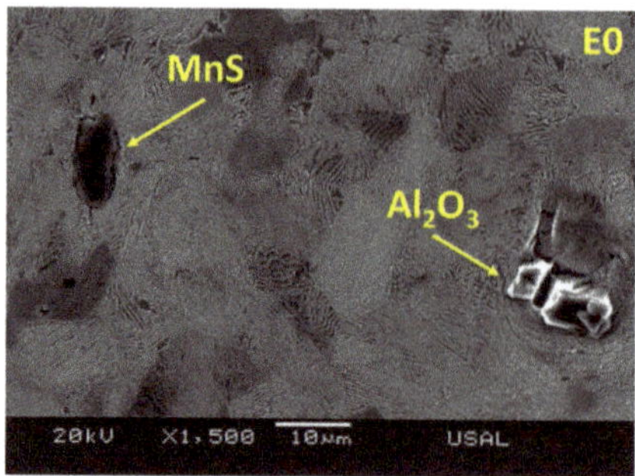

Figure 6. Second-phase particles in a hot rolled pearlitic steel.

Apart from the inclusions discussed above (those appearing more frequently), other types of inclusions were found (although more scarcely) used mainly to create new phases during the steel manufacturing. They were the following: (i) titanium oxides (probably in the form Ti$_2$O$_3$), (ii) manganese silicates (possibly of the type 2MnO.SiO$_2$, MnSiO$_3$), alumina silicates (SiO$_2$/AlO$_3$), (iii) titanium nitride (TiN) and (iv) vanadium nitride (VN). The main chemical elements appearing in an inclusion can be observed in the EDX spectrum given in Figure 7.

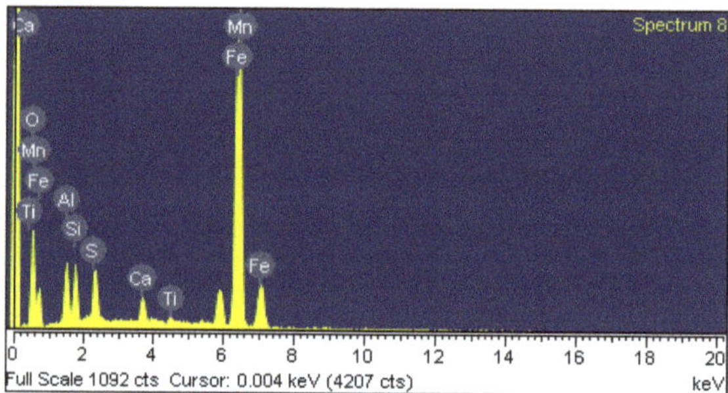

Figure 7. EDX spectrum of an inclusion containing several chemical elements.

The presence (in small amounts) of varying chemical elements (Ti, N and Ca), as well as other elements that did not appear in the metallographic analysis is not surprising. Titanium (Ti) is usually added to the steels during manufacturing, thereby producing Ti$_2$O$_3$ in form of small inclusions to enhance the formation of intergranular ferrite during the austenite-perlite transformation [51].

Nitrogen (N) is added to the steels to promote the formation of new phases from the austenite, thereby creating tiny particles of TiN. With regard to calcium (Ca), it is frequently added to the steel in certain amount (as SiCa powder) to diminish the level of sulfur (S) and, therefore, to diminish the volume fraction of sulfur and alumina inclusions in the steel [52].

4. Evolution of Inclusions with Cold Drawing

In the five families of steel (A, B, C, D and E) there is a common general trend in the matter of evolution with cold drawing of the inclusions. To analyze this, the existing inclusions in cold drawn pearlitic steels can be classified in two groups: (i) *sulphides* (MnS) able to undergo certain plastic strain and thus become deformed in the direction of cold drawing; (ii) inclusions consisting of *oxides and silicates* (of Al, Si, Fe, …), comparatively harder and more brittle, so that they can be fractured as the drawing degree (level of cumulative plastic strain) increases when the wire passes through the consecutive dies.

Figures 8–12 show several longitudinal metallographic sections of some wires studied in this work. The wire axis, or cold drawing direction, is represented by the vertical side of the micrographs in such figures. Figure 8 shows longitudinal metallographic sections of two hot rolled steels (E0 and A0) with zero cold drawing degree (i.e., that are not cold drawn at all). It is observed how the inclusions are perfectly adhered to the pearlitic metallic matrix surrounding them, although one was fractured during manufacturing by hot rolling.

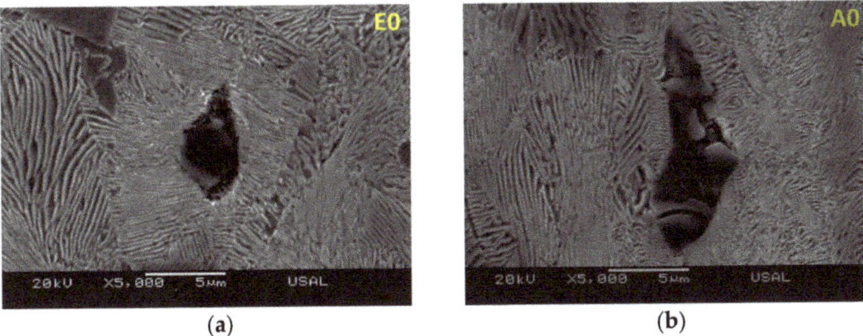

Figure 8. Inclusions in the longitudinal section of hot rolled pearlitic steels E0 and A0 (*zero degree of cold drawing*) showing *oxides* (**a**) and *sulfurs* (**b**).

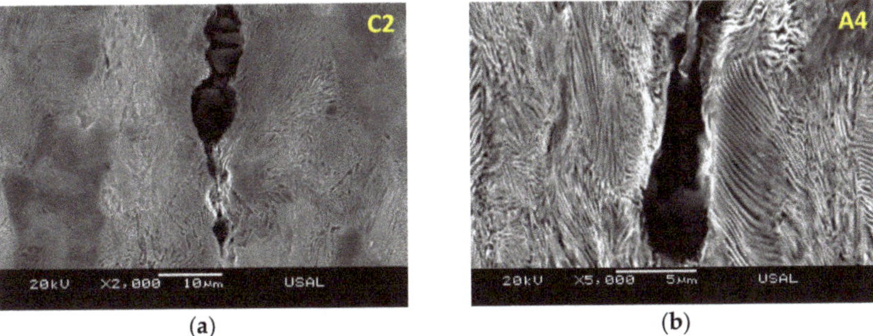

Figure 9. Inclusions in slightly drawn pearlitic steels C2 (with two drawing steps) and A4 (with four drawing steps). Principally MnS (**a**); complex oxide, mainly Al_2O_3 (**b**).

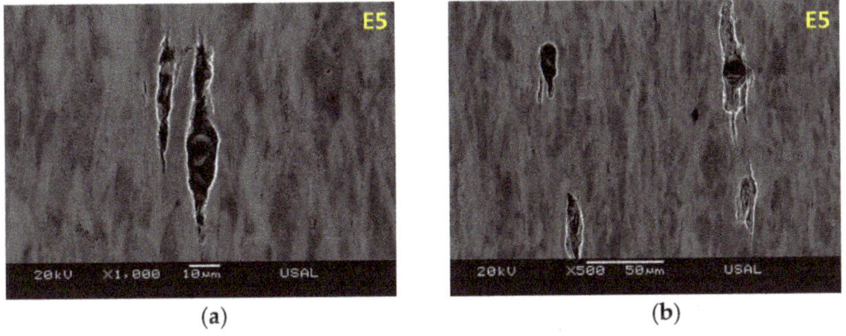

Figure 10. Inclusions in moderately drawn steel E5 (five drawing steps) at $1000\times$ (**a**) and $500\times$ (**b**).

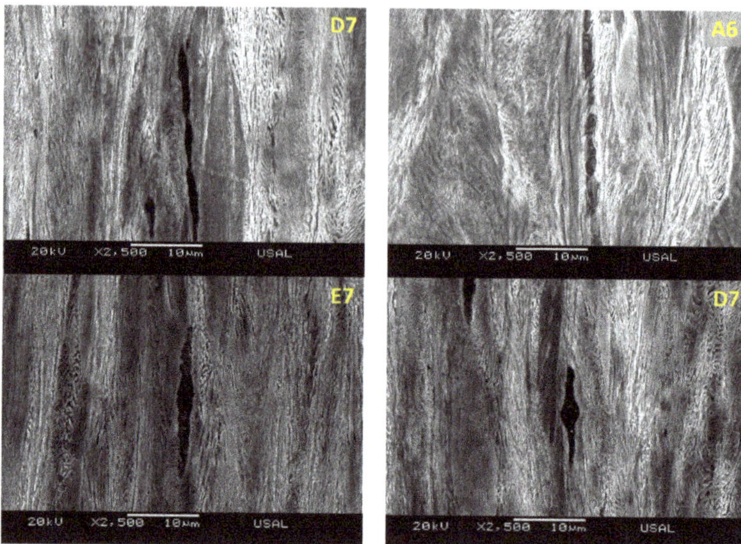

Figure 11. Axial micro-cracks generated by the presence of inclusions in heavily drawn pearlitic steels (commercial prestressing steel wires) (2500×).

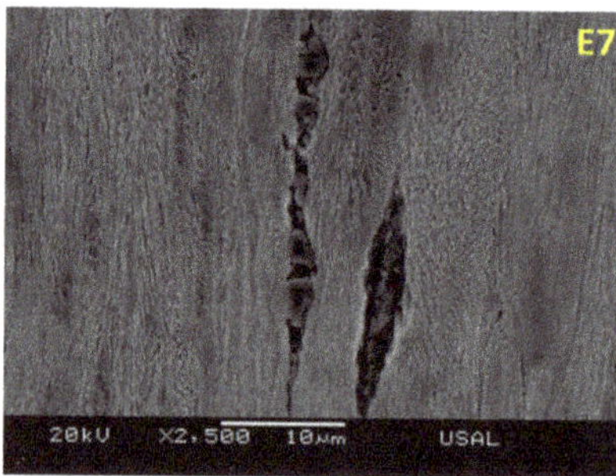

Figure 12. Axial micro-cracks generated in a heavily cold drawn pearlitic steel (commercial prestressing steel wire) created during the manufacturing process by cold drawing.

During manufacturing of commercial prestressing steel wires by progressive (multi-step) cold drawing, a strain hardening mechanism is activated in the pearlitic steels. During this process, the *inclusions* (second-phase particles) also evolve.

Figure 9 shows the microstructure of slightly cold drawn pearlitic steels C2 and A4 undergoing two and four cold drawing steps. It is observed how the inclusions present in the steels become fractured after passing through two and four dies (due to the transverse peripherical compression), thereby producing cracks at the interface between the inclusion and the matrix as a consequence of the plastic deformation undergone by the pearlitic matrix itself.

With regard to the differences between the five different families of the pearlitic steels under analysis, the steels of the group E are those containing greater and more numerous inclusions inside it. This provokes a higher density of microstructural defects (micro-defects) generated by the combined effect of heavy cold drawing and presence of many inclusions in the steel.

Figure 10 offers micrographs (axial or longitudinal cuts) of the microstructure of steel E5 (family E; five cold drawing steps) with two magnification levels of 1000× and 500×, where the frequent appearance of micro-defects (micro-cracks) generated by the presence of the inclusions can be observed. Such defects are potential initiators of macroscopic fracture of the cold drawn pearlitic steels, with loss of structural integrity.

In the matter of the most heavily cold drawn steel (commercial prestressing steel wire), obtained after passing of the initially hot rolled materials through the entire route of strain hardening (with the maximum number of cold drawing steps), micrographs of Figure 11 show evidence of micro-cracks aligned in the drawing direction (wire axis or longitudinal direction) since they are more slender than in the previous pearlitic steels (with a lower cold drawing degree). In Figure 11 the microstructure of steel families A, D and E is represented, observing that the inclusions are very fractured (they exhibit a lot of micro-damage) in their surrounding as a consequence of the high level of plastic strain undergone by the most heavily cold drawn pearlitic steels.

Figure 12 includes two parallel micro-cracks generated in the vicinity of inclusions appearing in a commercial prestressing steel wire E7 after heavy cold drawing. The inclusions exhibit evidence of previous fracturing during the manufacture process, and the micro-cracks appear in the close vicinity of the existing inclusions.

5. Role of Inclusions in the Fracture Behavior

From the results obtained in the metallographic analysis of the longitudinal sections of cold drawn pearlitic steels, a key question arises about the influence of inclusions on material (macro-) fracture behavior. To elucidate the relevant and fundamental question of whether (*or not*) the inclusions play a relevant role in fracture behavior of the steels, standard tension tests were performed on the different steel wires with a posterior (*postmortem*) fractographic analysis to elucidate the microscopic fracture modes.

In the case of hot rolled (not cold drawn at all) or slightly drawn pearlitic steels (corresponding to the first stages of the manufacturing chain with few passes through the dies) the fracture surface is contained in a plane perpendicular to the wire axis or cold drawing direction and exhibits a low-roughness appearance (Figure 13; left), i.e., it corresponds to an *isotropic fracture behavior*.

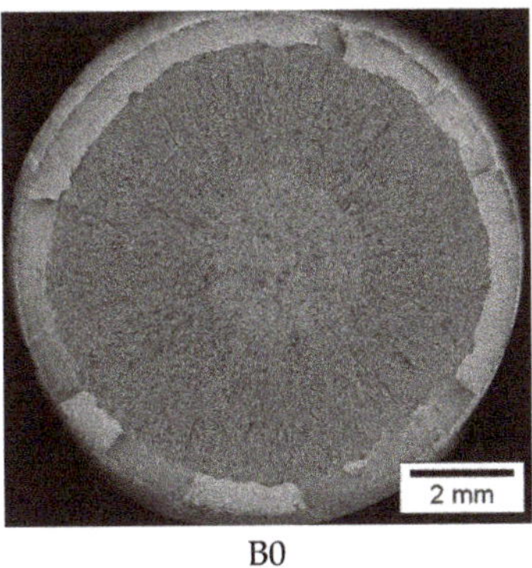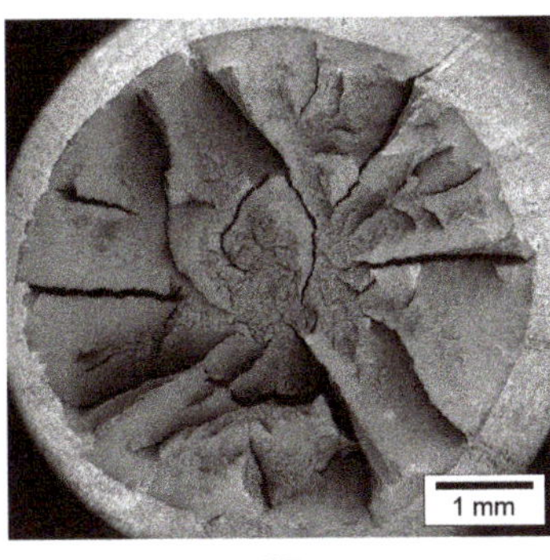

Figure 13. Fracture surfaces in a hot rolled bar (**left**) and a cold drawn wire (**right**).

On the other hand, in the case of heavily drawn pearlitic steels (associated with the last stages of the manufacturing chain with many passes through the dies) the fracture surface is *not* contained in a plane perpendicular to the wire axis or cold drawing direction and exhibits a high-roughness and irregular appearance (Figure 13; right), i.e., it corresponds to an *anisotropic fracture behavior* with frequent *local deflections* representing embryos of anisotropic fracture.

To understand such a disparity regarding fracture surface appearance, some microstructural features must be taken into account: (i) firstly, the oriented pearlitic microstructure after cold drawing in the matter of colonies and lamellae [22–25], (ii) secondly, the presence in the drawn steel of many zones with pre-damage (after heavy drawing) and even with micro-cracks aligned in the drawing direction and created in the vicinity of inclusions by debonding between the inclusion and the metallic matrix (as a consequence of the stress concentration created by the inclusion itself), as shown in Figure 14 in which many local regions with micro-cracks (aligned and oriented along the direction of cold drawing represented by the wire axis) may be observed.

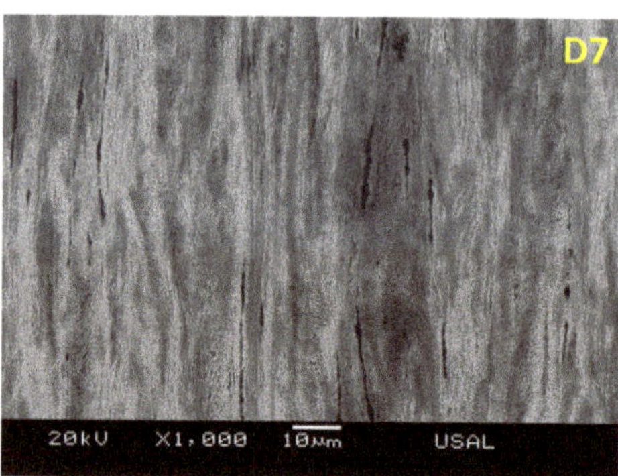

Figure 14. Multiple micro-cracks aligned in longitudinal direction (axial or cold-drawing direction, i.e., vertical side of the micrograph) in a cold drawn pearlitic steel.

During the standard tension tests performed on the wires, two main stages can be distinguished:

(i)   One first long phase (*before plastic instability*) during which material points are subjected to a uniform *uniaxial stress state*. This main stage lasts up to the maximum load point. During this phase the micro-cracks *do not* promote fracture since the main axial stress (in the direction of the wire) is parallel to the micro-cracks themselves.

(ii) One second shorter phase (*after plastic instability*) during which material points are subjected to a *triaxial stress state*. This second stage starts from the maximum load point. During this phase the micro-cracks *do* promote fracture since triaxial stress state has a hoop component that is perpendicular to the micro-cracks themselves.

An approximate classical solution for the triaxial stress state in a cylinder (bar or wire) after necking is due to Davidenkov and Spiridonova [53] as follows:

$$\frac{\sigma_r}{\sigma_Y} = \frac{a^2 - r^2}{2aR}; \frac{\sigma_z}{\sigma_Y} = 1 + \frac{a^2 - r^2}{2aR} \tag{2}$$

where σ_Y is the yield strength of the material, σ_z is the axial stress along the wire axis, σ_r is the radial stress (equal to the hoop stress σ_θ, due to the very small elastic strains in the neck compared with the plastic deformations, and due to the constant volume hypothesis which involves $\varepsilon_z = -2\varepsilon_r$), R is the curvature radius of the necking surface (similar to a blunt notch), a is the distance from the notch tip to the wire axis and r is the radial distance between the wire axis and the considered point (see Figure 15).

The curvature radius of the necking surface R was measure by means of a profile projector (NIKON V-12B, Tokyo, Japan). The average values obtained were 26.25 and 3.16 mm for the hot rolled steel E0 and the prestressing steel E7, respectively. With these values, and taking into account the Davidenkov and Spiridonova equations (2), it's possible to obtain the stress distribution (σ_z, σ_θ and σ_r; being $\sigma_\theta = \sigma_r$) along the wires net section in the previous instants to final fracture occurs. Such stress distribution is showed in Figure 16 for the case of the steels E0 and E7. The stress distribution is equal in both steels and for all considered variables, being maximum in the center (longitudinal wire axis, $r = 0$) and minimum in the periphery of the wires (deepest point in necking external surface, where $r = a$). The values are greater for the case of prestressing steel wires.

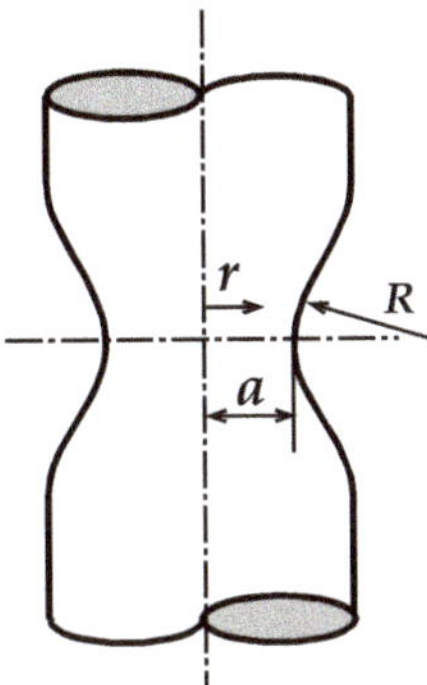

Figure 15. Scheme of the wire's necking during a tensile test.

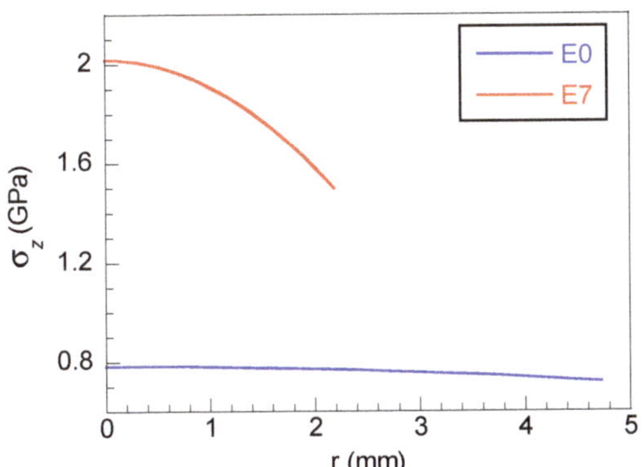
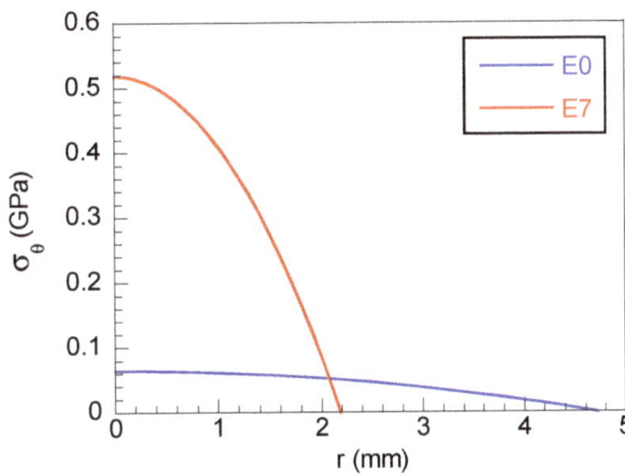

Figure 16. Stress distribution of σ_z (**left**) and σ_θ (**right**) along the transversal necking surface in steel E0 and E7.

Once the distribution of stresses along the minimum neck section is known, and considering that the shear stresses are null by symmetry, it is possible to obtain the stress triaxiality distribution along the radius r of the neck section. The stress triaxiality t (3) is the ratio between the hydrostatic stress σ_h (4) and the equivalent von Mises stress σ_{eq} (5).

$$t = \frac{\sigma_h}{\sigma_{eq}} \tag{3}$$

$$\sigma_h = \frac{1}{3}(\sigma_z + \sigma_\theta + \sigma_r) \tag{4}$$

$$\sigma_{eq} = \sqrt{\frac{1}{2}\left[(\sigma_z - \sigma_\theta)^2 + (\sigma_z - \sigma_r)^2 + (\sigma_\theta - \sigma_r)^2\right]} \tag{5}$$

Figure 17 (left) shows that the values of hydrostatic and equivalent stresses are higher for the case of prestressing steel (E7) in relation to the initial hot rolled bar (E0). The σ_h distribution shows a continuous decrement from the longitudinal wire axis ($r = 0$) till it reaches the external surface ($r = a$); that being said, the decrement is more pronounced in the prestressing steel E7. The equivalent von Mises stress σ_{eq} is constant along the radial coordinate for both steels.

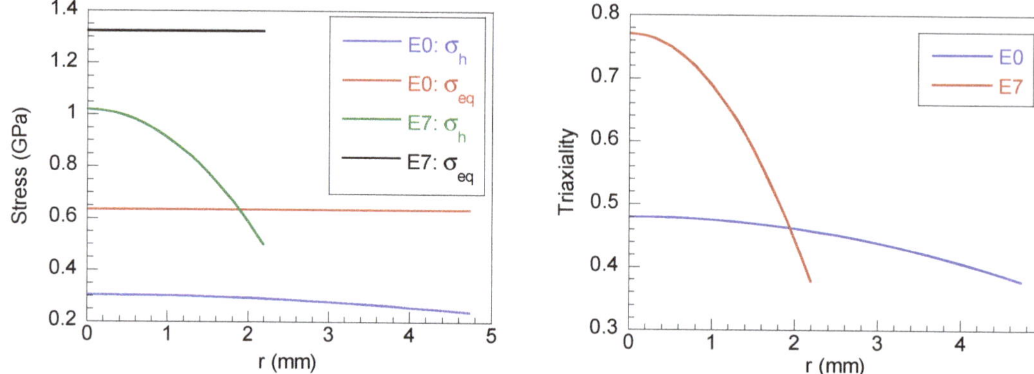

Figure 17. Distribution of hydrostatic stress (**left**), equivalent stress (**left**) and stress triaxiality (**right**) for the hot rolled steel E0 and the prestressing steel E7 along the necking section.

Figure 17 (right) shows the triaxiality factor distribution along the radial coordinate of the wire necking. The prestressing steel shows a higher level of triaxiality with respect to the hot rolled steel. In both steels the triaxiality factor is maximum in the longitudinal wire axis ($r = 0$) and minimum in the external surface of the neck.

Heavily cold drawn steels (Figure 18) exhibit *locally anisotropic fracture behavior* due to the presence of multiple micro-cracks oriented and aligned along the drawing direction (created by debonding between the inclusions and the pearlitic matrix) and the triaxial stress state (with radial and hoop components) generated after necking. Both factors are relevant to create an anisotropic fracture in heavily cold drawn pearlitic steels in which there is a coexistence of:

(i)   Elevated values of σ_Y, σ_R and σ_θ (see Figures 3, 4 and 16).
(ii)   High level of stress triaxiality (with $\sigma_z > \sigma_\theta$), as indicated in Figure 17.
(iii) Many micro-cracks created from the inclusions, aligned and oriented along the drawing axis (Figure 14). The hoop stress σ_θ during plastic instability (*necking*) before final fracture induces *locally anisotropic fracture* with crack deflections and appearance of crests and valleys.

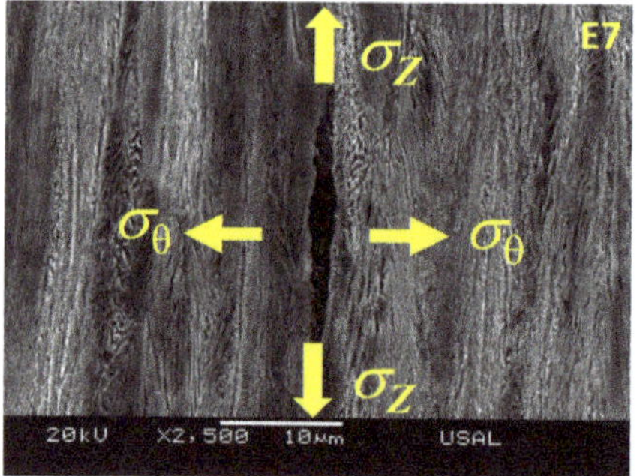

Figure 18. Anisotropic fracture behavior in a cold drawn pearlitic steel (E7).

Hot rolled (not cold drawn at all) and slightly drawn pearlitic steels (Figure 19) exhibit an *isotropic fracture behavior* due to the absence of the afore-said micro-cracks. In this case the stress state is defined by:

(i)   Moderated values of σ_Y, σ_R and σ_θ (see Figures 3, 4 and 16).
(ii)   Lower level of stress triaxiality (with $\sigma_z > \sigma_\theta$), as showed in Figure 17.

(iii) Scarcity (or practical absence) of pre-damage and micro-cracks. Therefore, the triaxial stress state generated during necking is unable to produce local deflections and thus locally anisotropic fracture.

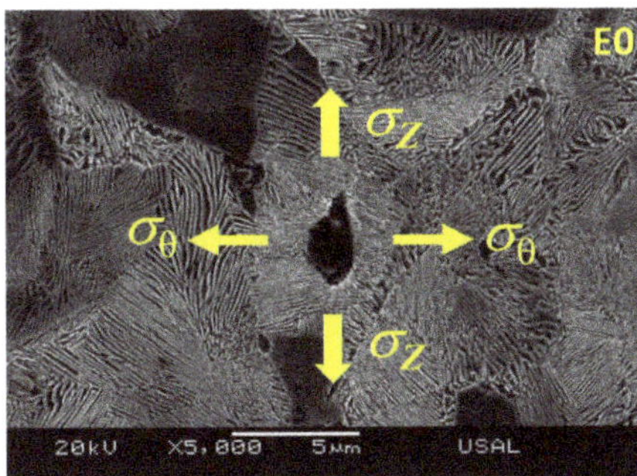

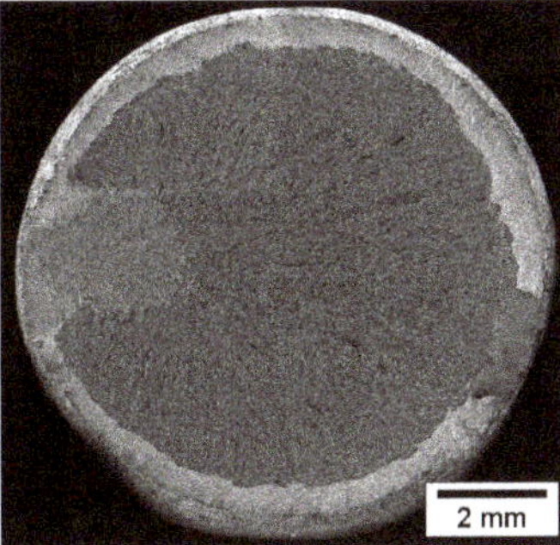

Figure 19. Isotropic fracture behavior in a hot rolled pearlitic steel (E0).

6. Conclusions

The anisotropic fracture behavior of cold drawn pearlitic steel wires subjected to standard tension tests can be explained on the basis of the triaxial stress state generated during necking, together with the presence of many aligned and oriented micro-cracks generated during the cold drawing process.

Such micro-cracks appear in the vicinity of second-phase particles (*inclusions*) in the course of manufacturing of prestressing steel wires by progressive cold drawing after debonding between the inclusion and the pearlitic matrix due to the differences with regard to plastic behavior between the matrix and the inclusion itself.

Author Contributions: J.T. conceived and designed the research; F.-J.A. and B.G. performed the microscopy analysis, J.T., F.-J.A. and B.G. analyzed the data and wrote the paper. All authors have read and agreed to the published version of the manuscript.

Funding: This research was funded by the following Spanish Institutions: Ministry for Science and Technology (MICYT; Grant MAT2002-01831), Ministry for Education and Science (MEC; Grant BIA2005-08965), Ministry for Science and Innovation (MICINN; Grant BIA2008-06810), Ministry for Economy and Competitiveness (MINECO; Grant BIA2011-27870) and Junta de Castilla y León (JCyL; Grants SA067A05, SA111A07, SA039A08 and SA132G18).

Institutional Review Board Statement: Not applicable.

Informed Consent Statement: Not applicable.

Data Availability Statement: Not applicable.

Conflicts of Interest: The authors declare no conflict of interest. The funders had no role in the design of the study; in the collection, analyses, or interpretation of data; in the writing of the manuscript, and in the decision to publish the results.

References

1. Gil-Sevillano, J. Cleavage-limited maximum strength of work-hardened B.C.C. polycrystals. *Acta Metall.* **1986**, *34*, 1473–1485.
2. Toribio, J. On the intrinsic character of the stress-strain curve of a prestressing steel. *J. Testing Eval.* **1992**, *20*, 357–362.
3. Toribio, J. Cold drawn eutectoid pearlitic steel wires as high performance materials in structural engineering. *Struct. Integr. Health Monit.* **2006**, *2*, 239–247.
4. Toribio, J.; Ayaso, F.J.; González, B.; Matos, J.C.; Vergara, D.; Lorenzo, M. Tensile fracture behavior of progressively-drawn pearlitic steels. *Metals* **2016**, *6*, 114. [CrossRef]
5. Borchers, C.H.; Kirchheim, R. Cold-drawn pearlitic steel wires. *Prog. Mater. Sci.* **2016**, *82*, 405–444. [CrossRef]
6. Toribio, J. Structural integrity of progressively cold-drawn pearlitic steels: From Raffaello Sanzio to Vincent van Gogh. *Proced. Struct. Integr.* **2017**, *3*, 3–10. [CrossRef]

7. Toribio, J.; González, B.; Matos, J.C.; Kharin, V. Evaluation by sharp indentation of anisotropic plastic behaviour in progressively drawn pearlitic steel. *ISIJ Int.* **2011**, *51*, 843–848. [CrossRef]
8. Toribio, J.; Ovejero, E.; Toledano, M. Microstructural bases of anisotropic fracture behaviour of heavily drawn steel. *Int. J. Fract.* **1997**, *87*, L83–L88.
9. Toribio, J.; Ayaso, F.J. Fracture performance of progressively drawn pearlitic steel under triaxial stress states. *Mater. Sci.* **2001**, *37*, 707–717. [CrossRef]
10. Toribio, J.; Ayaso, F.J. Investigation of the type of cleavage related to anisotropic fracture in heavily drawn steels. *J. Mater. Sci. Lett.* **2002**, *21*, 1509–1512. [CrossRef]
11. Toribio, J.; Ayaso, F.J. Micromechanics of fracture in notched samples of heavily drawn steel. *Int. J. Fract.* **2002**, *115*, L29–L34. [CrossRef]
12. Toribio, J.; Ayaso, F.J. Anisotropic fracture behavior of cold drawn steel: A materials science approach. *Mater. Sci. Eng.* **2003**, *343*, 265–272. [CrossRef]
13. Toribio, J.; Ayaso, F.J. Image analysis of exfoliation fracture in cold drawn steel. *Mater. Sci. Eng.* **2004**, *387*, 438–441. [CrossRef]
14. Toribio, J.; González, B.; Matos, J.C. Cleavage stress required to produce fracture path deflection in cold-drawn prestressing steel wires. *Int. J. Fract.* **2007**, *144*, 189–196. [CrossRef]
15. Toribio, J.; González, B.; Matos, J.C. Strength anisotropy and mixed mode fracture in heavily drawn pearlitic steel. *Fatigue Fract. Eng. Mater. Struct.* **2013**, *36*, 1178–1186. [CrossRef]
16. Tanaka, M.; Saito, H.; Yasumaru, M.; Higashida, K. Nature of delamination cracks in pearlitic steels. *Scr. Mater.* **2016**, *112*, 32–36. [CrossRef]
17. He, Y.; Xiang, S.; Shi, W.; Liu, J.; Ji, X.; Yu, W. Effect of microstructure evolution on anisotropic fracture behaviors of cold drawing pearlitic steels. *Mater. Sci. Eng.* **2017**, *683*, 153–163. [CrossRef]
18. Toribio, J. Anisotropy of hydrogen embrittlement in cold drawn pearlitic steel: A tribute to Mantegna. *Proced. Struct. Integr.* **2020**, *28*, 2438–2443. [CrossRef]
19. Embury, J.D.; Fisher, R.M. The structure and properties of drawn pearlite. *Acta Metall.* **1966**, *14*, 147–159. [CrossRef]
20. Langford, G. Deformation of pearlite. *Metall. Trans.* **1977**, *8*, 861–875. [CrossRef]
21. Ridley, N. A review of the data on the interlamellar spacing of pearlite. *Metall. Trans.* **1984**, *15*, 1019–1036. [CrossRef]
22. Toribio, J.; Ovejero, E. Microstructure evolution in a pearlitic steel subjected to progressive plastic deformation. *Mater. Sci. Eng.* **1997**, *234*, 579–582. [CrossRef]
23. Toribio, J.; Ovejero, E. Microstructure orientation in a pearlitic steel subjected to progressive plastic deformation. *J. Mater. Sci. Lett.* **1998**, *17*, 1037–1040. [CrossRef]
24. Toribio, J.; Ovejero, E. Effect of cumulative cold drawing on the pearlite interlamellar spacing in eutectoid steel. *Scr. Mater.* **1998**, *39*, 323–328. [CrossRef]
25. Toribio, J.; Ovejero, E. Effect of cold drawing on microstructure and corrosion performance of high-strength steel. *Mech. Time Depend. Mater.* **1998**, *1*, 307–319. [CrossRef]
26. Languillaume, J.; Kapelski, G.; Baudelet, B. Cementite dissolution in heavily cold drawn pearlitic steel wires. *Acta Mater.* **1997**, *45*, 1201–1212. [CrossRef]
27. Zelin, M. Microstructure evolution in pearlitic steels during wire drawing. *Acta Mater.* **2002**, *50*, 4431–4447. [CrossRef]
28. Hall, E.O. The deformation and ageing of mild steel: III Discussion of results. *Proc. Phys. Soc.* **1951**, *64*, 747–753. [CrossRef]
29. Petch, N.J. The cleavage strength of polycrystals. *J. Iron Steel Inst.* **1953**, *174*, 25–30.
30. Choi, H.C.; Park, K.T. The effect of carbon content on the Hall-Petch parameter in the cold drawn hypereutectoid steels. *Scr. Mater.* **1996**, *34*, 857–862. [CrossRef]
31. Nam, W.J.; Bae, C.M.; Lee, C.S. Effect of carbon content on the Hall-Petch parameter in cold drawn pearlitic steel wires. *J. Mater. Sci.* **2002**, *37*, 2243–2249. [CrossRef]
32. Toribio, J. Relationship between microstructure and strength in eutectoid steels. *Mater. Sci. Eng.* **2004**, *387*, 227–230. [CrossRef]
33. Toribio, J.; González, B.; Matos, J.C. Microstructure and mechanical properties in progressively drawn pearlitic steel. *Mater. Trans.* **2014**, *55*, 93–98. [CrossRef]
34. Kang, S.K.; Gow, K.W. An improved metallographic technique for the study of nonmetallic inclusions in steel. *Metallography* **1978**, *11*, 219–222. [CrossRef]
35. Roberts, W.; Lehtinen, B. An in situ SEM study of void development around inclusions in steel during plastic deformation. *Acta Metall.* **1975**, *24*, 745–758. [CrossRef]
36. Goodwin, S.J.; Noble, F.N.; Eyre, B.L. Inclusion nucleated ductile fracture in stainless steel. *Acta Metall.* **1989**, *37*, 1389–1398. [CrossRef]
37. Everett, R.K.; Geltmacher, A.B. Spatial distribution of MnS inclusions in HY-100 steel. *Scr. Mater.* **1999**, *40*, 567–571. [CrossRef]
38. Raghupathy, V.P.; Srinivasan, V. Determination of volume fraction of sulfide inclusions in steels. *Metallography* **1981**, *14*, 87–97. [CrossRef]
39. Eid, N.M.A.; Thomason, P.F. The nucleation of fatigue cracks in a low-alloy steel under high-cycle fatigue conditions and uniaxial loading. *Acta Metall.* **1979**, *27*, 1239–1249. [CrossRef]
40. Fowler, G.J. The influence of non-metallic inclusions on the threshold behavior in fatigue. *Mater. Sci. Eng.* **1979**, *39*, 121–126. [CrossRef]

41. Chapetti, M.D.; Tagawa, T.; Miyata, T. Ultra-long cycle fatigue of high-strength carbon steels. Part I: Review and analysis of the mechanism of failure. *Mater. Sci. Eng.* **2003**, *356*, 227–235. [CrossRef]
42. Chapetti, M.D.; Tagawa, T.; Miyata, T. Ultra-long cycle fatigue of high-strength carbon steels. Part II: Estimation of fatigue limit for failure from internal inclusions. *Mater. Sci. Eng.* **2003**, *356*, 236–244. [CrossRef]
43. You, D.; Michelic, S.K.; Presoly, P.; Liu, J.; Bernhard, C.H. Modeling inclusion formation during solidification of steel: A review. *Metals* **2017**, *7*, 460. [CrossRef]
44. Guo, J.; Yang, W.; Shi, X.; Zheng, Z.; Liu, S.; Duan, S.; Wu, J.; Guo, H. Effect of sulfur content on the properties and MnS morphologies of DH36 structural steel. *Metals* **2018**, *8*, 945. [CrossRef]
45. Ånmark, N.; Karasev, A.; Jönsson, P.G. The effect of different non-metallic inclusions on the machinability of steels. *Materials* **2015**, *8*, 751–783. [CrossRef]
46. Yan, W.; Chen, W.; Li, J. Quality control of high carbon steel for steel wires. *Materials* **2019**, *12*, 846. [CrossRef]
47. Schäfer, B.J.; Sonnweber-Ribic, P.; Hassan, H.; Hartmaier, A. Micromechanical modeling of fatigue crack nucleation around non-metallic inclusions in martensitic high-strength steels. *Metals* **2019**, *9*, 1258. [CrossRef]
48. Liu, H.; Hu, D.; Fu, J. Analysis of MnS inclusions formation in resulphurised steel via modeling and experiments. *Materials* **2019**, *12*, 2028. [CrossRef] [PubMed]
49. Qiu, G.; Zhan, D.; Li, C.; Yang, Y.; Qi, M.; Jiang, Z.; Zhang, H. Influence of inclusions on the mechanical properties of RAFM steels via Y and Ti addition. *Metals* **2019**, *9*, 851. [CrossRef]
50. Yang, W.; Peng, K.; Zhang, L.; Ren, Q. Deformation and fracture of non-metallic inclusions in steel at different temperatures. *J. Mater. Res. Technol.* **2020**, *9*, 15016–15022. [CrossRef]
51. Shim, J.H.; Oh, Y.J.; Suh, J.Y.; Cho, Y.W.; Shim, J.D.; Byum, J.S.; Lee, D.N. Ferrite nucleation potency of non-metallic inclusions in medium carbon steels. *Acta Mater.* **2001**, *49*, 2115–2122. [CrossRef]
52. Wilson, A.D. The influence of thickness and rolling ratio on the inclusion behavior in plate steels. *Metallography* **1979**, *12*, 233–255. [CrossRef]
53. Kachanov, L.M. *Fundamentals of the Theory of Plasticity*; Dover Publications: Mineola, NY, USA, 2004; pp. 311–314.

Article

Influence of Uniaxial Deformation on Texture Evolution and Anisotropy of 3104 Al Sheet with Different Initial Microstructure

Sofia Papadopoulou [1,2,*], **Vasilis Loukadakis** [2], **Zisimos Zacharopoulos** [2] and **Spyros Papaefthymiou** [2]

[1] ELKEME S.A., 61st km Athens-Lamia Nat. Road, Oinofyta, 32011 Viotia, Greece
[2] Laboratory of Physical Metallurgy, Division of Metallurgy and Materials, School of Mining & Metallurgical Engineering, National Technical University of Athens, 9, Her. Polytechniou Str., Zografos, 15780 Athens, Greece; vasilis.loukadakis@gmail.com (V.L.); szacharopoulos@hotmail.com (Z.Z.); spapaef@metal.ntua.gr (S.P.)
* Correspondence: spapadopoulou@elkeme.vionet.gr; Tel.: +30-2262-60-4309

Abstract: Optimum mechanical behavior is achieved by means of controlling microstructural anisotropy. The latter is directly related to the crystallographic texture and is considerably affected by thermal and mechanical processes. Therefore, understanding the underlying mechanisms relating to its evolution during thermomechanical processing is of major importance. Towards that direction, an attempt to identify possible correlations among significant microstructural parameters relating to texture response during deformation was made. For this purpose, a 3104 aluminum alloy sheet sample (0.5 mm) was examined in the following states: (a) cold rolled (with 90% reduction), (b) recovered and (c) fully recrystallized. Texture, anisotropy as well as the mechanical properties of the samples from each condition were examined. Afterwards, samples were subjected to uniaxial loading (tensile testing) while the most deformed yet representative areas near the fractured surfaces were selected for further texture analysis. Electron backscatter diffraction (EBSD) scans and respective measurements were conducted in all three tensile test directions (0°, 45° and 90° towards rolling direction (RD)) by means of which the evolution of the texture components, their correlation with the three selected directions as well as the resulting anisotropy were highlighted. In the case of the cold-rolled and the recovered sample, the total count of S2 and S3 components did not change prior to and after tensile testing at 0° towards RD; however, the S2 and S3 sum mostly consisted of S3 components after tensile testing whereas it mostly consisted of S2 components prior to tensile testing. In addition, the aforementioned state was accompanied by a strong brass component. The preservation of an increased amount of S components, and the presence of strain-free elongated grains along with the coexistence of a complex and resistant-to-crack-propagation substructure consisting of both high-angle grain boundaries (HAGBs) and subgrain boundaries (SGBs) led into an optimal combination of Δr and r_m parameters.

Keywords: 3104 alloy; crystallographic texture; cold rolling; tensile testing

Citation: Papadopoulou, S.; Loukadakis, V.; Zacharopoulos, Z.; Papaefthymiou, S. Influence of Uniaxial Deformation on Texture Evolution and Anisotropy of 3104 Al Sheet with Different Initial Microstructure. *Metals* **2021**, *11*, 1729. https://doi.org/10.3390/met11111729

Academic Editor: Nikki Stanford

Received: 28 September 2021
Accepted: 27 October 2021
Published: 29 October 2021

Publisher's Note: MDPI stays neutral with regard to jurisdictional claims in published maps and institutional affiliations.

1. Introduction

Beverage cans and food packaging are manufactured by use of 3xxx alloy series such as 3004 and 3104 due to their high strength and formability [1]. After many years of research and optimization of the relevant thermomechanical processes, the industry heavily relies on the use of those alloys [2]. Although these alloy series cover current market demands, the existence of failures along with a relatively high percentage of rejected material, are frequently observed during the forming processes in which those products are subjected [2,3]. This phenomenon is attributed to sheet anisotropy provoking the formation of "ears" of varying extent and orientation. Earing is a dominant defect induced during the deep drawing process and defined as the creation of waviness on the uppermost portion of the resulting cup. Rolling texture components favor earing principally at 45°/135° with respect to the RD [4,5]. With regards to earing, a low ear ratio can be achieved through an optimal combination of crystals with an orientation that results in both 45°/135° and 0°/90° ears [6–9]. The crystallographic components which are formed during recrystallization depend on the stacking fault energy (SFE), the previous deformation state, the annealing temperature as well as the precipitation state. Metals with high SFE (like Al) mainly exhibit a deformation substructure consisting of dislocation tangles and cells instead of banded, linear arrays of dislocations [10]. The anisotropy is, essentially, a result of the texture obtained throughout the thermomechanical processes., A thorough understanding of the microstructural evolution, may result into the effective control of the resulting texture after each thermomechanical process. This essentially relates to the response to deformation with the aim of minimizing rejections, increasing the product's life cycle, and, consequently, improving the Al industry's economic and environmental footprint.

Moreover, study of the texture components which are present at the most susceptible to fracture areas can provide fruitful information in relation to the underlying failure mechanisms. Grain orientation may strongly affect the deformed microstructures as metals, in general, are plastically deformed due to crystal lattice shifts in planes where more space is available. Furthermore, areas within a grain can be separated from each other due to the glide of dislocations, thus, plastic distortions are imported in the crystal lattice [11]. Various studies [12–15] have highlighted a strong correlation between the texture of aluminum alloys, e.g., β-fiber percentage, and their mechanical properties, formability as well as strain rate sensitivity. In particular, the presence of a cube texture tends to worsen the mechanical properties of the material, whereas mixed textures may improve them [16].

The present article aims to underline the effect of uniaxial tensile loading on texture for several thermomechanical states of a 3104 Al cold-rolled sheet (deformed, recovered, and recrystallized). In addition, the response of the microstructural characteristics such as the misorientation grain boundaries is studied to examine whether a correlation among the abovementioned characteristics and anisotropy exists. The effect of different initial microstructures, and, in particular the strong effect of the uniaxial loading is also addressed. Study of the behavior of the different texture components and subsequent substructures under uniaxial loading through the proposed experimental approach is critical and innovative since it could lead an enhanced understanding of the texture evolution. This information could then be used for the induction of the desirable amount of texture components to be able to produce Al sheet with optimum response towards drawing. Towards that direction the texture near the "neck" area of all tested tensile specimens was examined. The obtained results were reviewed and discussed regarding the microstructural characteristics such as the misorientation grain boundaries and r-values. The correlation among the texture evolution and the resulting anisotropy is crucial since studies which combine texture and plastic deformation are of great practical relevance.

2. Materials and Methods

The AA 3104 alloy sheet samples (Table 1) that were examined were produced through direct chill (DC) casting and originated from a 600 mm thickness plate. Additionally, the sheet samples were subjected to hot and cold rolling as well as to thermal treatments.at As such, the final samples that were extracted were representative of both the recovered and the fully recrystallized states. Regarding sample designation, a four-letter system was used for the nomenclature as follows: The two first letters indicate the sheet sample state where c.r. stands for cold rolled, rc. stands for recovered, rx. for recrystallization, AR for as-received state, RD for samples subjected to tensile parallel to the rolling direction (RD), 45 for 45° towards RD, TR for transverse to RD. Various annealing temperatures were applied through the range of 200 °C to 340 °C with a 10 h ramp and a 2 h soaking time similar to common industrial practice. Through this process indicative samples were selected for the two aforementioned states. It is noted that sample selection was based on the use of grain orientation spread (GOS) maps, through which the states under examination were detected.

Table 1. Samples description.

State	Condition	Sample ID
Cold rolling 90% reduction (deformed)	As received state	c.r.AR
	After tensile testing of //RD sample	c.r.RD
	After tensile testing of 45° towards RD sample	c.r.45
	After tensile testing of ⊥RD sample	c.r.TR
Recovery annealing (recovered)	As received state	rc.AR
	After tensile testing of //RD sample	rc.RD
	After tensile testing of 45° towards RD sample	rc.45
	After tensile testing of ⊥RD sample	rc.TR
Recrystallization annealing (fully recrystallized)	As received state	rx.AR
	After tensile testing of //RD sample	rx.RD
	After tensile testing of 45° towards RD sample	rx.45
	After tensile testing of ⊥RD sample	rx.TR

2.1. Tensile Testing

The materials were subjected to uniaxial tensile strength, as per the requirements of the ISO 6892-1 testing standard. In particular, the R_p (yield strength—YS), R_m (ultimate tensile strength—UTS) and A_g (elongation) parameters were measured. The tensile strength specimens were cut, in orientations of 0°, 45° and 90° relevant to the RD. The planar anisotropy parameter Δr (1) and the normal anisotropy parameter r_m or r-bar (2) were measured according to [17]:

$$\Delta r = \frac{r_0 + r_{90} - 2r_{45}}{2} \tag{1}$$

$$r_m = \frac{r_0 + r_{90} + 2r_{45}}{4} \tag{2}$$

where r_0, r_{45}, and r_{90} are the r-values in orientations of 0°, 45° and 90° relevant to the RDn, respectively.

2.2. Hardness and Conductivity Measurements

The Vickers method was used for hardness testing. A load of 0.2 kgf was selected for hardness measurements whereas for the electrical conductivity measurements, a frequency of 240 kHz was used. The mean values presented originate from the average of 10 measurements.

2.3. Optical Microscopy

Samples were sectioned parallel to the RD and were cold mounted to avoid any non-desirable annealing effects. Metallographic preparation was conducted through successive grinding and polishing steps, whereas the grain structure was revealed through the use of Barker's electrolytic etching according to references [3,18]. Examination of the grain structure was performed with polarized light and by use of a Nikon Epiphot 300 inverted metallographic microscope (Nikon, Minato City, Tokyo, Japan). Optical metallographic examination was conducted at a magnification range of 500×–1000× after Barkers etching.

2.4. Metallographic Examination—Texture

Microscopic analysis was performed using a JEOL IT-800 HL Scanning Electron Microscope (SEM, JEOL Ltd., Akishima, Tokyo, Japan) under a 20 kV accelerating voltage, coupled with an EDAX Apollo XF, equivalent to octane super EDS, silicon drift detector (SDD, EDAX, Mahwah, NJ, USA) in conjunction with TEAM (EDAX, Mahwah, NJ, USA) and SMILE VIEW™ Map software (JEOL Ltd., Akishima, Tokyo, Japancity, and Digital Surf, Besançon, France). EBSD analysis was carried out using an EDAX Hikari XP EBSD, high-speed camera (EDAX camera, EDEN Instruments SAS, Valence Cedex, France). When a polycrystalline metallic material, is subjected to thermomechanical processing, a dominant orientation, namely texture is observed. The texture could be formed through the deformation, annealing, or as a result of both processes. Texture can be categorized into several components, represented by {hkl}<uvw>. The latter are further divided into two groups, consisting of deformation and annealing texture components.

Calculation of the various texture components, resulting from the applied manufacturing process, was performed according to [19]:

- *Rolling components*: Copper {112}<111>, S2{213}<-1-42>, S3 (or S) {123}<634>, Taylor{4411}<11118>, Brass {011}<211>
- *Recrystallization components*: Cube {001}<100>, P{011}<111>, Q{013}<2-31>, R{124}<211>, Goss{011}<100>

EBSD scans were collected through a hexagonal grid. For the EBSD analysis, samples were examined in the as polished condition with a tilt of $70°$ relative to the electron beam. Voltage was set at 20 kV, while the step size was set at 0.1 μm to 1 μm depending on the material condition and the magnification used ($100\times$–$2000\times$) (examined areas 30×30 μm^2–150×150 μm^2). Regarding the tensile samples, the areas near the "neck" were further investigated metallographically after the tests. To characterize the fractured samples, EBSD scans at low magnification were also performed. Through this examination, the area exhibiting the higher kernel average misorientation (KAM) values and the higher amount of subgrain boundaries was selected for further analysis at higher magnification. The black points observed in the inverse pole figure (IPF) maps are points with of low confidence index and were therefore excluded from the measurements. Scanning data were then further analyzed by means of a post-processing software package (OIM) for the various texture components to be identified and calculated. The misorientation angles were determined according to $2°$–$5°$ (subgrain angle boundaries), $5°$–$15°$ (low-angle boundaries) and $15°$–$62.8°$ (HAGBs). It is noted that misorientation angles less than $0.5°$ corresponding to dense dislocation, are not considered reliable for evaluation purposes due to the angular resolution of the EBSD [20]. Finally, post-processing was conducted to clean up the data (grain dilation).

3. Results

3.1. Mechanical Behavior (Tensile Testing, Anisotropy)

The lower Δr (-0.20) and highest rm values (1.44) were recorded in the case of the rc. AR sample (rc.RD, rc.45 and rc.TR). In Figure 1 the effect of thermal treatment to ultimate tensile strength (UTS) is presented. As expected, the specimens that were tempered in a higher temperature appear to have inferior strength, due to their recrystallized microstructure. Likewise, these samples, are more ductile than the c.r.AR sample which exhibited elongation values lower than 5%. An increase of the electrical conductivity (EC) was evident and could be attributed to the precipitation of dispersoids.

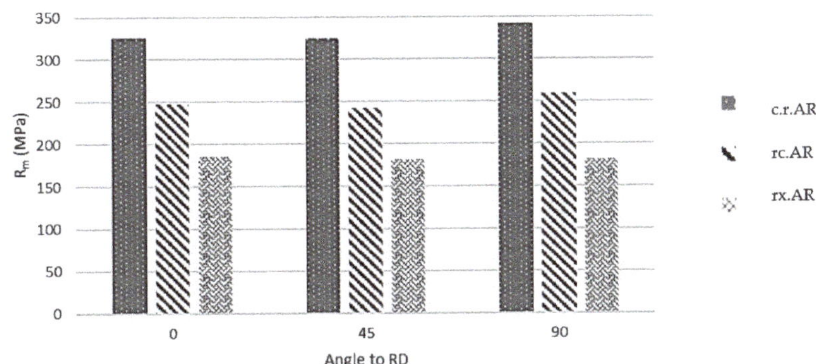

Figure 1. Tensile strength of the samples depending on their relative angle to RD.

The planar anisotropy parameter (Δr) is an indicator of the ability of a material to demonstrate a "non-earing" behavior (lower values are preferable) while r_m, is an excellent indicator of the ability of a material to be subjected to deep drawing (higher is better), (Table 2). The maximum load before fracture was recorded first samples transverse to the RD direction (c.r.TR and rc.TR). After recrystallization, the maximum load values were recorded in the parallel to RD direction (1102 N). A similar behavior was observed for R_m while the $Rp_{0.2}$ values at 90° relative to RD, were the highest among all measurements (Table 2, Figures 1 and 2). Elongation did not exhibit any noteworthy correlation with the direction examined; however, it was observed that after each thermal treatment (recovery and recrystallization annealing) the ductility almost doubled from ≈4% to ≈7% and, eventually, increased to 18%. An inversely proportional correlation was observed in hardness and conductivity measurements, however (Figure 3).

Table 2. Mechanical properties of the examined samples.

Sample ID	Max Load (N)	$Rp_{0.2}$ (Mpa)	R_m (Mpa)	A_g (%)	Δr	r_m	Conductivity (IACS%)	Hardness $HV_{0.2}$
c.r.RD	1932.5 ± 2.1	300 ± 5.7	325 ± 0.0	4 ± 1.4				
c.r.45	1932.5 ± 2.1	288 ± 15.6	324.5 ±0.7	3 ± 0.0	0.05	0.88	38.3	107
c.r.TR	2034 ± 1.4	302 ± 2.8	342 ± 0.0	4 ± 1.4				
rc.RD	1469 ± 1.4	209.5 ± 3.5	247 ± 0.0	7.5 ± 0.7				
rc.45	1443.5 ± 4.9	221.5 ± 2.1	242.5 ± 0.7	7.5 ± 0.7	−0.20	1.44	38.9	83
rc.TR	1541 ± 5.7	233.5 ± 2.1	259 ± 1.4	5.5 ± 0.7				
rx.RD	1102 ± 1.4	70.5 ± 9.2	185 ± 0.0	17.5 ± 07				
rx.45	1077 ± 2.8	70 ± 7.1	181 ± 0.0	18 ± 1.4	−0.14	1.05	39.8	51
rx.TR	1076.5 ± 6.0	75 ± 1.4	181 ± 3.5	18.5 ± 0.0				

The recovered sample exhibited an isotropic behavior with a maximum deviation of the mechanical properties for each sample close to 10% (Figure 4). With regards to orientations 0°, 45° and 90°, it is noted that the most pronounced shift of mechanical properties during recrystallization occurred in the 90° orientation (YS: from 234 MPa to 75 MPa and UTS: from 259 MPa to 181 MPa) and therefore can be characterized as a more sensitive direction, with a UTS reduction of the order of 76 MPa. In the 0° direction (YS: from 210 MPa to 76 MPa and UTS: from 247 MPa to 188 MPa) a decrease of UTS by 60 MPa is noted, whereas in the 45° direction (YS: from 222 MPa to 75 MPa and UTS: from 243 MPa at 185 MPa) a decrease in UTS of 58 MPa is recorded. Bases on the aforementioned results, it is noted that as the angle increases from $0°$ to $45°$ stress decreases whereas in $90°$ it increases again to reach its maximum value.

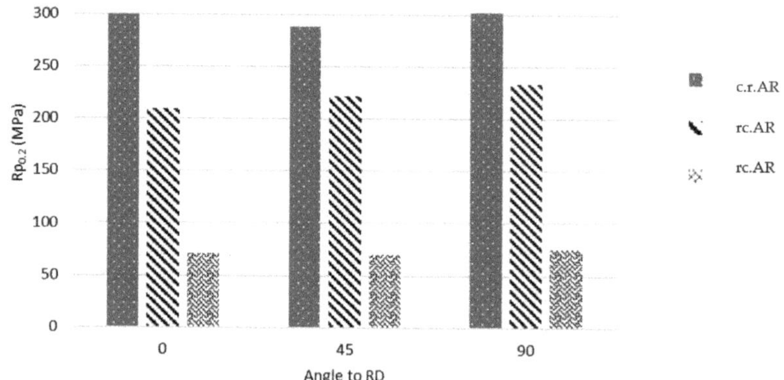

Figure 2. Yield strength of the samples depending on their relative angle to RD.

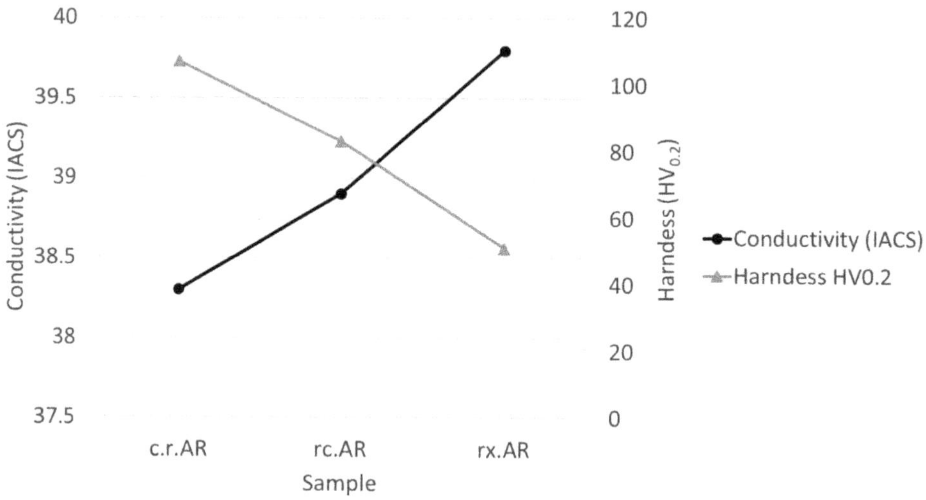

Figure 3. Conductivity vs. Hardness diagram of the three states examined (cold-rolled, recovered, and recrystallized).

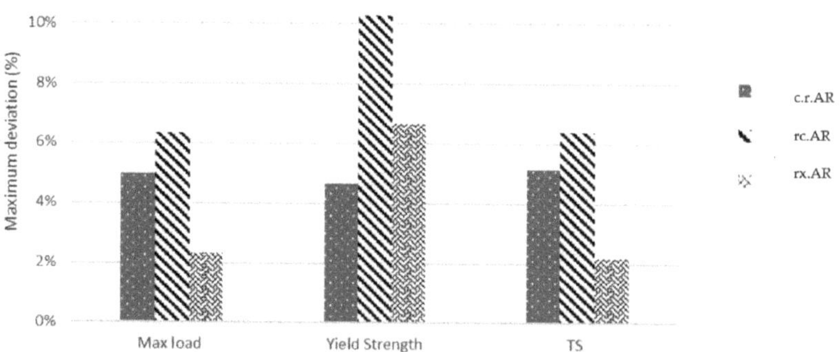

Figure 4. Maximum deviation of the mechanical properties for each sample.

Regarding the planar anisotropy parameters, it is noted that the predominant deformation texture in the case of the rc.AR sample is equal to 1.5 which is the highest among the examined annealing temperatures (Figure 5). It is also mentioned that in the case of rc.AR sample, the calculated Δr is equal to -0.2. As such a weaker "earing" phenomenon is expected.

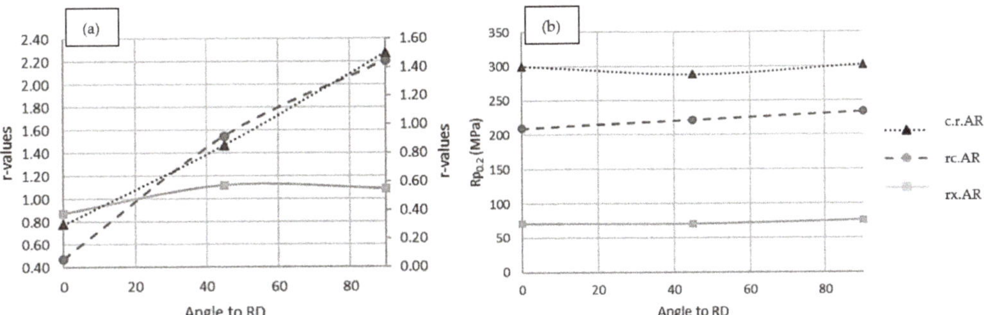

Figure 5. (**a**) Plastic strain ratio (r value) of examined sheet samples, (**b**) in-plane anisotropy in a yield strength, $R_{p0.2}$.

The cold-rolled sample exhibited the maximum recorded $R_{p0.2}$, R_m, and hardness values compared to the other samples. After the recovery annealing process, all measured parameters decreased except for elongation which almost doubled compared to the previous state. At the fully recrystallized state, the sheet sample exhibited the lowest measured $R_{p0.2}$ and R_m values as well as the maximum elongation. The measured r-values led to the conclusion that the optimal combination of the resulting properties along with a low possibility of failure during drawing was exhibited by the recovered sample exhibiting $\Delta r = 0.20$ and $r_m = 1.44$. The hardness value decreased during annealing due to the underlying strain relief. At the same time, electrical conductivity exhibited a slight increase indicating a "cleaner" matrix.

3.2. Microstructural Analysis and Fractography

Microstructural evolution of the c.r.AR sample is presented in Figure 6a. It starts with a strongly deformed microstructure (fibrous), and retains its morphology after recovery/annealing. The recrystallization annealing lead to a fully recrystallized microstructure, consisting of coarse equiaxed grains throughout its thickness (Figure 6c).

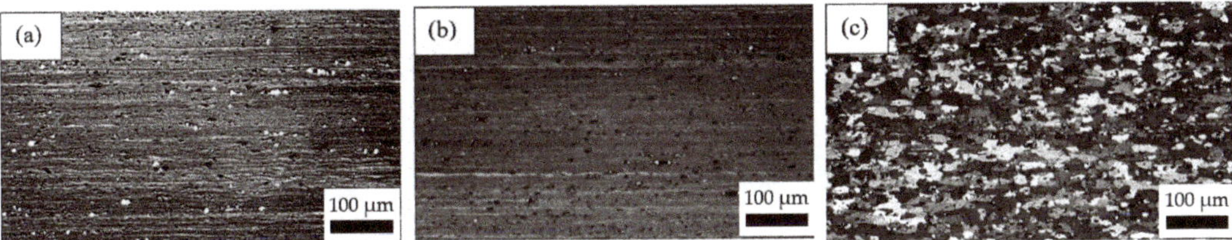

Figure 6. Optical micrographs after Barkers electrolytic etching showing the microstructure of (**a**) c.r.AR, (**b**) rc.AR and (**c**) rx.AR.

Metallographic and fractographic inspection of all samples after tensile testing was also performed. An indicative micrograph of sample rx.TR is shown in Figure 7. The directivity of the equiaxed grains due to uniaxial loading, is noted and indicated by the arrows.

Figure 7. Indicative optical micrograph after Barkers etching of sample rx.TR. The black arrow indicated the tensile load direction.

The fracture surfaces of the broken tensile specimens were examined by use of SEM. The numerous deep and large dimples observed highlight the ductile nature of the tested material (Figure 8a). An indicative 3D reconstruction of the observed surface is presented below in Figure 8b to provide a quantitative perspective of the observed dimples. All samples exhibited similar fracture features.

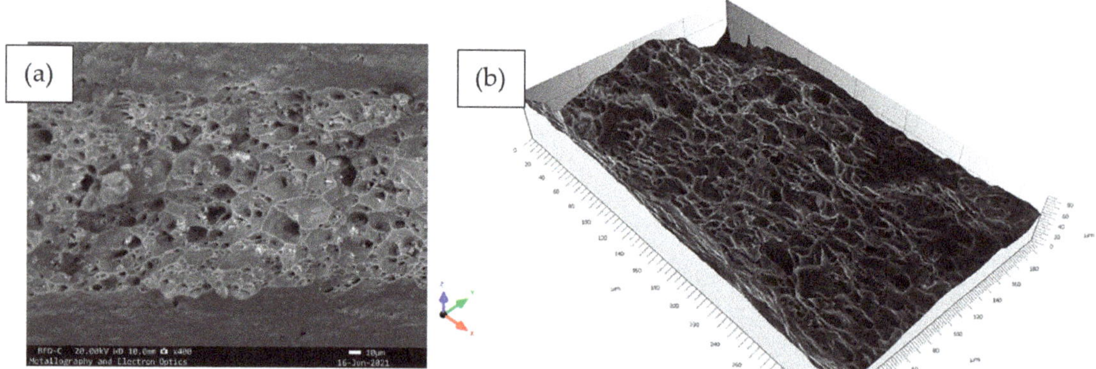

Figure 8. (**a**) SEM fractography and (**b**) indicative 3D reconstruction of surface topography from one or several SEM fractography of sample rx.TR.

3.3. EBSD Scans and Analysis

After the metallographic examination of the samples, EBSD scans were conducted on the fractured samples near the "neck" area. It is noted that due to tensile loading many voids were formed near this area. At the same time, the deformation led to the distortion of the lattice, thus many areas cannot be evaluated with regards to texture. Those points were excluded from post-processing examination due to their low confidence index (CI) value and are marked with black color in all maps (see Figures 9–12). An indicative EBSD scan, which was used to detect the most appropriate area is presented in Figure 9. The fracture surface is shown in Figure 9a where the prevalence of the (001), (111) and (112) orientations is evident in the IPF map. In Figure 9b the (white) arrows indicate the areas where the highest KAM values were observed, which is also the area where the maximum density of SGBs (Figure 9c) was measured. In the final map of this figure (Figure 9d) the coexistence of both rolling and recrystallization texture components is underlined in the selected area. For the main part of the EBSD data, a comparative map of the samples in the as-received and after-tensile-testing conditions is constructed for all examined orientations. GOS maps were used for the selection of the most indicative samples from all three examined states (deformed, recovered, and recrystallized), (Table 3). The IPF intensity did not exhibit a high variance indicating a non-strong overall texture.

Table 3. Results of misorientation angles, IPF intensities, and GOS percentages.

Sample ID	Condition	IPF Intensity	GOS	
c.r.AR	Deformed	2.7	0°–1°	3.7
			1°–10°	95.1
rc.AR	Recovered	2.4	0°–1°	18.7
			1°–10°	66.6
rx.AR	Recrystallized	2.0	0°–1°	73.3
			1°–10°	26.1

As previously observed through the optical micrographs, the sample c.r.AR exhibited a fibrous, highly deformed microstructure with elongated grains (Figure 10a). After tensile testing, voids were formed, and the resulting texture was significantly influenced. According to Table 4, KAM values increased from 0.3 in the initial state to ≈0.5 after tensile testing in all examined orientations, due to the augmentation of disorders within the grains. The highest KAM value was measured for the cr.TR sample. Regarding the misorientation angles, the highest value recorded was that of low-angle grain boundaries (LAGBs) for all states. After tensile testing, a significant amount of LAGBs was present (≈20%), as expected. The texture components of the cr.AR sample were dominated by rolling texture components. S2 accounted for ≈40% of the identifiable components whereas it was reduced to 10.6%, 1.2% and 0% for samples cr.RD, c.r.45 and c.r.TR respectively, after tensile testing. S3 (usually referred as S), in the cold-rolled state was 12.5% and after uniaxial loading a maximum value of 43.4% was obtained for the sample parallel to the RD whereas it was almost absent in all other orientations. The sum of S2 and S3 before and after tensile testing remained stable at approximately 50% of the overall texture (the brass component exhibited a strong presence and along with the S component, they reached about 60% of the overall texture. The behavior of P and Goss texture components is noteworthy, in the sense that they could barely be detected prior to tensile testing whereas in sample c.r.45 they exhibited an increase of 15.1% and 64.2% respectively, after testing. Augmentation of the P component was detected in the case of sample c.r.TR also, reaching 44.4% of the overall texture. The presence of the R component is also pronounced, and increased in the orientation parallel to the RD whereas it could not be detected in all other orientations. An opposing behavior was observed with regards to Goss components, which are barely detected = in all states except for the c.r.45 sample. The differences between orientations can be explained through an alteration of the IPF plot at the bottom of Figure 9 where the initial state at (101) rotating towards (111) and (112) orientations can be seen. The intensity of the IPFs did not exhibit any significant differences. In the case of the c.r.AR sample the predominant recrystallization texture component is the R at 10.4%, while the predominant crystallographic deformation texture component is the S accounting for more than 40% of the overall texture (Table 4).

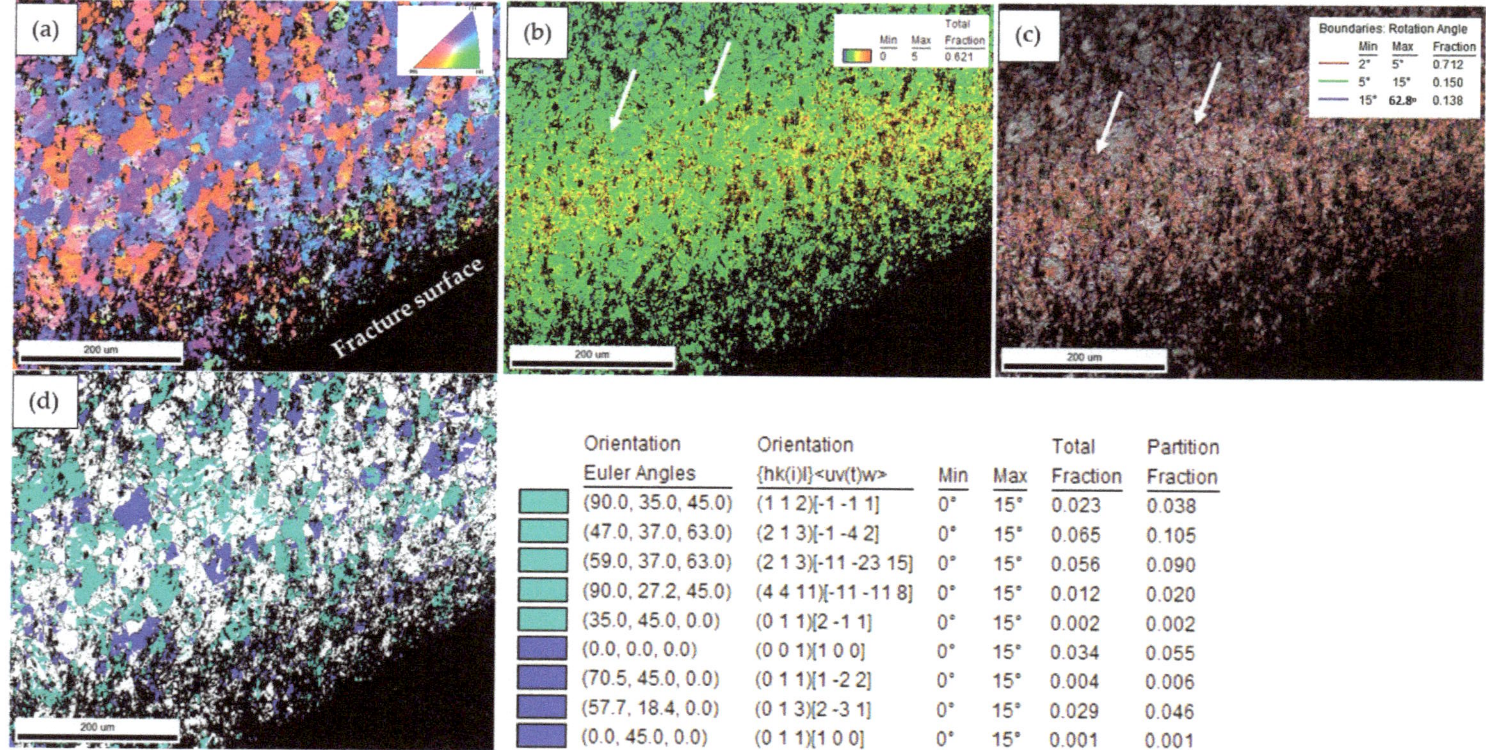

	Orientation Euler Angles	Orientation {hk(i)l}<uv(t)w>	Min	Max	Total Fraction	Partition Fraction
	(90.0, 35.0, 45.0)	(1 1 2)[-1 -1 1]	0°	15°	0.023	0.038
	(47.0, 37.0, 63.0)	(2 1 3)[-1 -4 2]	0°	15°	0.065	0.105
	(59.0, 37.0, 63.0)	(2 1 3)[-11 -23 15]	0°	15°	0.056	0.090
	(90.0, 27.2, 45.0)	(4 4 11)[-11 -11 8]	0°	15°	0.012	0.020
	(35.0, 45.0, 0.0)	(0 1 1)[2 -1 1]	0°	15°	0.002	0.002
	(0.0, 0.0, 0.0)	(0 0 1)[1 0 0]	0°	15°	0.034	0.055
	(70.5, 45.0, 0.0)	(0 1 1)[1 -2 2]	0°	15°	0.004	0.006
	(57.7, 18.4, 0.0)	(0 1 3)[2 -3 1]	0°	15°	0.029	0.046
	(0.0, 45.0, 0.0)	(0 1 1)[1 0 0]	0°	15°	0.001	0.001

Figure 9. Indicative scan of sample rx.RD which was accomplished at lower magnification to detect the areas of interest (indicated by the black arrow). (**a**) IPF map, (**b**) KAM map, (**c**) misorientation angle map and (**d**) map illustrating the crystallographic components (light blue: rolling components and dark blue: recrystallization components).

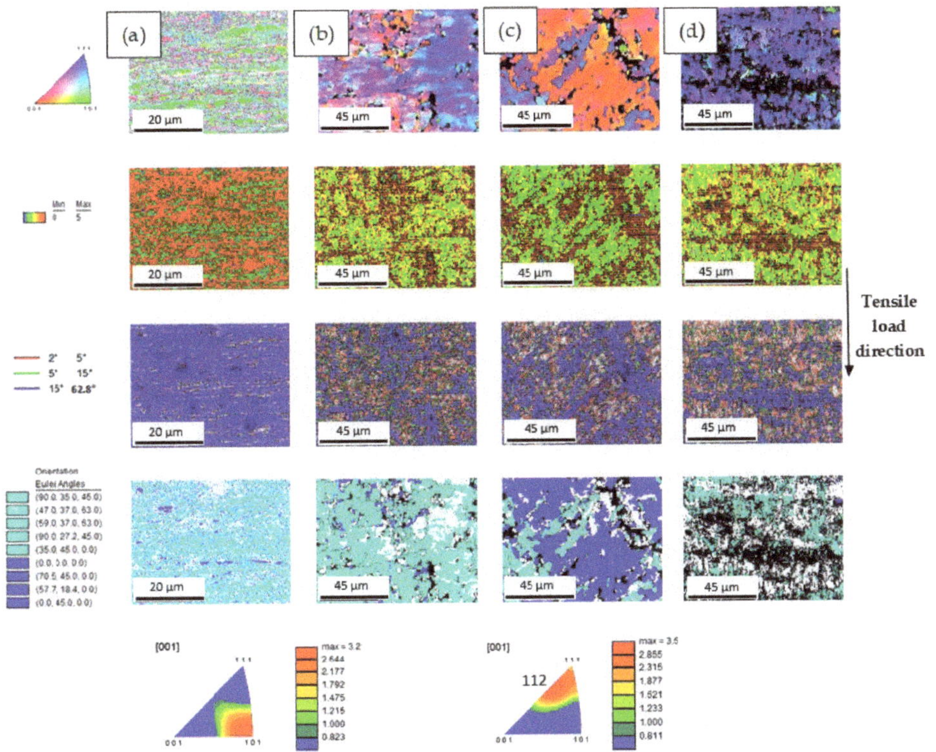

Figure 10. EBSD scans and respective maps for sample (**a**) column CRAR, (**b**) CRPAR, (**c**) CR45 and (**d**) CRTRA.

Table 4. KAM values, misorientation, and texture components of the cold-rolled sheet sample.

	c.r.AR	c.r.RD	c.r.45	c.r.TR
KAM values	0.3	0.5	0.4	0.6
Misorientation angles				
2–5°	**69.8**	**70.6**	77.7	**70.0**
5–15°	9.2	23.5	15.9	23.3
15–62.5°	21.0	5.9	6.4	6.7
Rolling components				
Copper (90, 35, 45)	1.1	12.2	0.4	0.5
S2 (47, 37, 63)	**40.4**	10.6	1.2	0
S3 (59, 37, 63)	**12.5**	43.4	0.2	0.2
Taylor/Dillamore (90, 27, 45)	**0.3**	**1.8**	0.5	0.1
Brass (35, 45, 90)	**3.4**	**21.7**	0.1	0.1
Recrystallization components				
Cube (0, 0, 0)	**1.9**	0.5	0	0
P (70, 45, 0)	**2.3**	0.9	15.1	44.4
Q (58, 18, 0)	**4.2**	**2.0**	12.1	6.1
R (57, 29, 63)	**10.4**	5.8	1.2	
Goss (0, 45, 0)	**1.6**	2.6	64.2	1.4

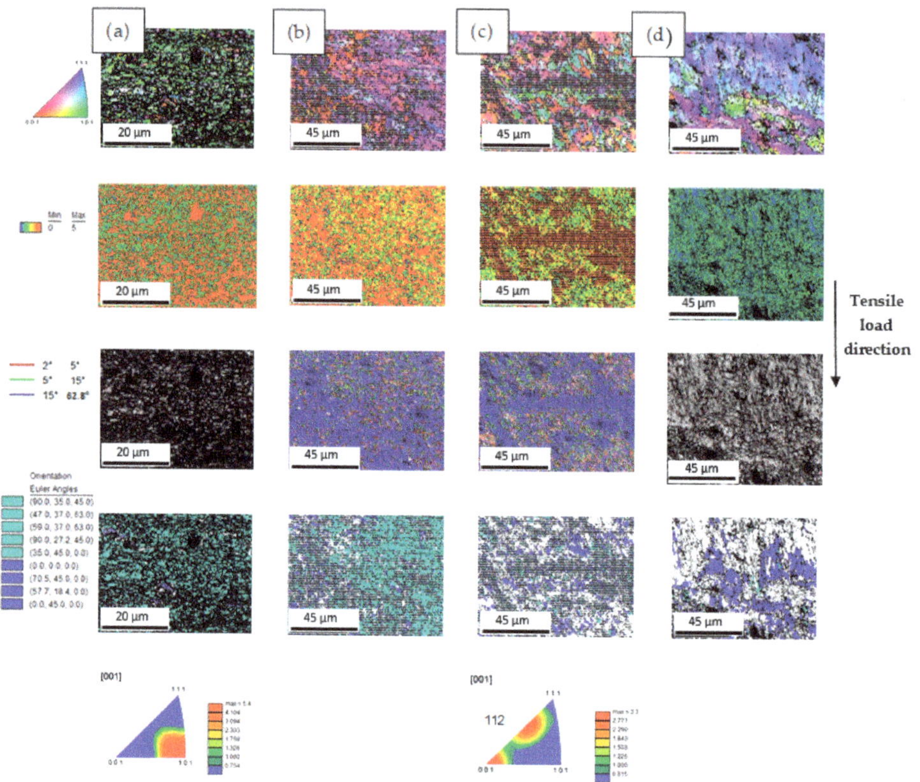

Figure 11. EBSD scans and respective maps for sample (**a**) rc.AR, (**b**) rc.RD, (**c**) rc.45 and (**d**) rc.TR.

The cold-rolled sample exhibited a high amount of SGBs accompanied by a high percentage of S2, S3, and R components. After tensile testing, the SGBs preserved their high percentage in the overall texture whereas the previously mentioned crystallographic components (S2, S3 and R) were accompanied by a quite noticeable increase of brass and copper components for the sample parallel to the RD. At 45° relative to the RD, P, Q, and mainly Goss components dominated the observed texture, reaching altogether approximately 92% of the overall texture. The P component exhibited an increase of 45% for the sample transverse to RD.

The sheet sample (after being recovered) did not exhibit any noteworthy differences with regard to grain morphology whereas it exhibited a higher variation of KAM values compared to the c.r.AR sample in all examined orientations (Figure 11 and Table 5). The maximum KAM value after tensile testing was measured in the orientation of 90° (1.0), and exhibited behavior similar to the sample c.r.AR. As for the misorientation grain boundaries, the recovered sample at the as-received state exhibited a higher amount of HAGBs (57%) compared to the cold-rolled condition, along with a significant amount of SGBs l (35.9%). After tensile testing, an increase of the subgrain boundaries (SGBs) and a decrease of the HAGBs is observed, with a maximum of ≈73% being recorded for the rc.TR sample. In the case of sample c.r.AR, the S2 component exhibited a maximum intensity in the as-received condition and was succeeded by S3 after tensile testing in 0° relative to the RD. Both S2 and S3 amounted for 50% of the overall texture. After recovery, the opposite behavior is observed, where the S3 component prevails at as-received state. Again, the sum of these two rolling components, amounts for 50% of the overall texture in the as-received state as well as after tensile testing in 0°. The behavior of the brass and cube components is also noteworthy since, brass decreases from 30% to 12% in sample rc.RD, while cube increases from 1.5% to 5.6%. An analogous, to the previous sample, presence of P and Goss components, as well as Q is detected. P and Goss increased in sample rc.45 (from 0.3% and 0.5% to 9.4% and 5.7% respectively) whereas Q increased in the case of sample rc.TR (from 1.1% to 19.4%). The R component continued to be the dominant in the final recrystallization texture components, accounting for 10.4% of the overall texture for the rc.TR sample and an even higher percentage of 33.4% after tensile testing in the rc.RD sample whereas in the other direction, it was vanished. The dominant (101) orientations were retained after the annealing process whereas after the deformation process that tendency altered since the grains were more prone to rotate towards the (112) and (001) orientations. The (101) orientation of the crystal is parallel to the assigned sample direction while after tensile testing a double fiber is provoked, namely the (001) and (112).

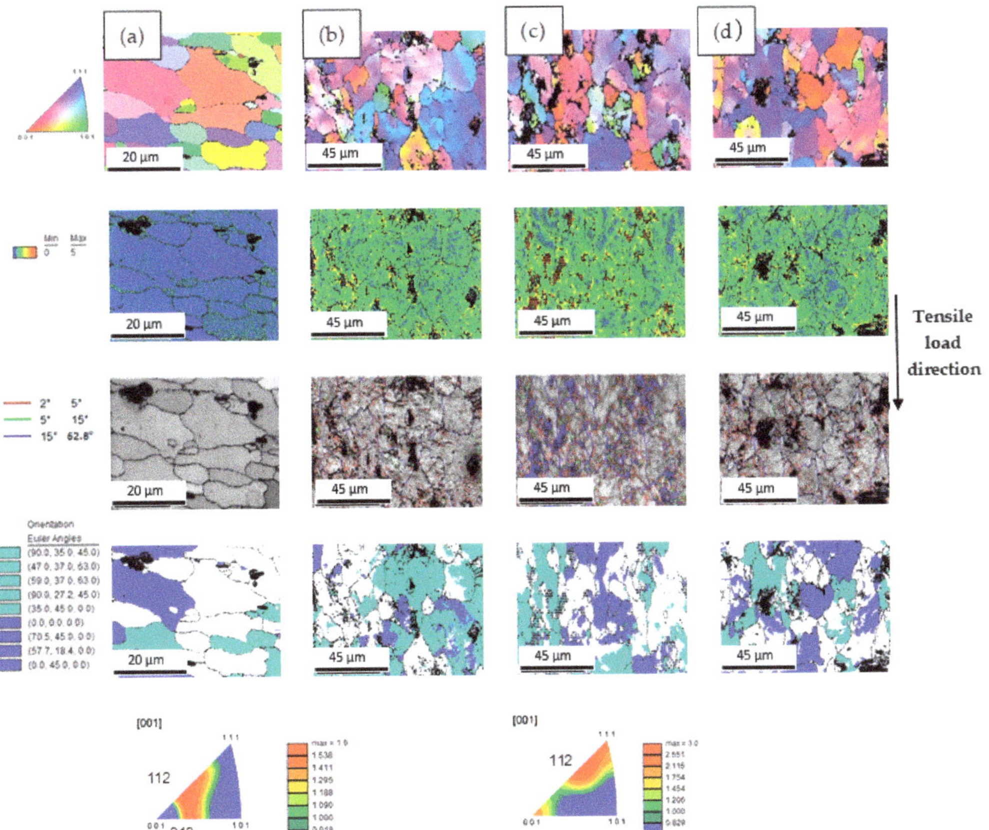

Figure 12. EBSD scans and respective maps for sample (**a**) rx.AR, (**b**) rx. RD, (**c**) rx.45 and (**d**) rx.TR.

For the final examined state, the recrystallized, rx.AR sample, exhibited a low KAM value (Table 6). HAGBs dominated the rx.AR samples since it mainly consisted of well-formed equiaxed recrystallized strain-free grains, a condition that was not preserved after tensile testing. In 0° and 90° textures were dominated by SGBs whereas in 45° HAGBs were mostly evident. In addition, the copper component decreased, whereas S2, S3, Taylor, and brass increased. P increased significantly in 90° while Q increased in all directions with the maximum being observed in the 45° orientation. The amount of Goss was reduced from ≈24% to ≤6% for all orientations. Finally, the initial state exhibited a dominance of

(112) and (313) orientations whereas after the deformation process the (001), (112), and (111) were the most prominent orientations that were observed. A weaker texture according to IPFs was observed in the as-received condition whereas a 50% increase was observed after tensile testing.

Table 5. KAM values, misorientation, and texture components of the recovered sheet sample.

	rc.AR	rc.RD	rc.45	rc.TR
KAM values	0.1	0.8	0.9	1.0
Misorientation angles				
2–5°	35.9	**51.9**	**43.8**	**72.6**
5–15°	7.1	28.3	29.1	16.6
15–62.5°	**57.0**	19.8	27.1	10.7
Rolling components				
Copper (90, 35, 45)	40.8	8.3	0.7	0.6
S2 (47, 37, 63)	**6.5**	**15.6**	2.7	0.4
S3(59, 37, 63)	**45.2**	**31.6**	1.3	0.2
Taylor/Dillamore (90, 27, 45)	**2.6**	**2.4**	0.7	0.2
Brass (35, 45, 90)	**29.5**	**11.9**	0.5	0.5
Recrystallization components				
Cube (0, 0, 0)	**1.5**	**5.6**	1.2	1.4
P (70, 45, 0)	**0.3**	**0.6**	9.4	7.2
Q (58, 18, 0)	**1.1**	**3.9**	2.9	19.4
R (57, 29, 63)	**5.6**	**6.1**	0	0.4
Goss (0, 45, 0)	**0.5**	**2.3**	5.7	0.1

Table 6. KAM values, misorientation, and texture components of the recrystallized sheet sample.

	rx.AR	rx.RD	rx.45	rx.TR
KAM values	0.1	1.0	0.8	1.0
Misorientation angles				
2–5°	3.4	**70.4**	39.4	**63.7**
5–15°	4.6	8.1	11.5	8.3
15–62.5°	**92.0**	21.6	**49.1**	28.0
Rolling components				
Copper (90, 35, 45)	6.4	3.6	5.1	5.8
S2 (47, 37, 63)	**22.0**	3.0	11.3	2.9
S3(59, 37, 63)	**1.4**	15.9	5.8	9.4
Taylor/Dillamore (90, 27, 45)	**1.1**	13.2	2.8	2.5
Brass (35, 45, 90)	**8.5**	12.1	0.5	2.8
Recrystallization components				
Cube (0, 0, 0)	**0.1**	**1.3**	0.4	9.1
P (70, 45, 0)	**2.6**	**0.8**	3.7	10.8
Q (58, 18, 0)	**8.2**	**8.4**	10.0	8.8
R (57, 29, 63)	**0.1**	**6.7**	4.3	4.8
Goss (0, 45, 0)	**23.6**	**0.0**	5.6	0.8

The highest tensile strength was measured in the case of the cold-rolled sample where a rotation from (101) to (112) and (111) is also noted. The recovered sample initiated with (101) as well and rotated mainly towards (112) and to a lesser extend to (001) orientations. The recrystallized sample exhibited (112) and (313) orientations which were partly retained (112), (001) and (111). The cold-rolled sample and the recovered sample, did not exhibit significant differences with regards to texture and the dominant orientations. A different behavior however was observed after tensile testing due to the lack of the effect of dislocations inside the recovered grains. The main orientations rotation observed was (101) to (112). This shift resulted into a higher stiffness of the examined sheet samples in compared to the recrystallized sample where no significant orientation shift was noted (the (112) orientation was present before and after tensile).

The texture components are qualitatively represented by the orientation distribution functions (ODF) maps below. Although IPFs did not exhibit a strong texture in some cases, ODFs highlight the presence of the crystallographic components and their solid presence for each sample (Figure 13). The black arrows indicate the positions of the components, with an intensity high enough to be measured and evaluated.

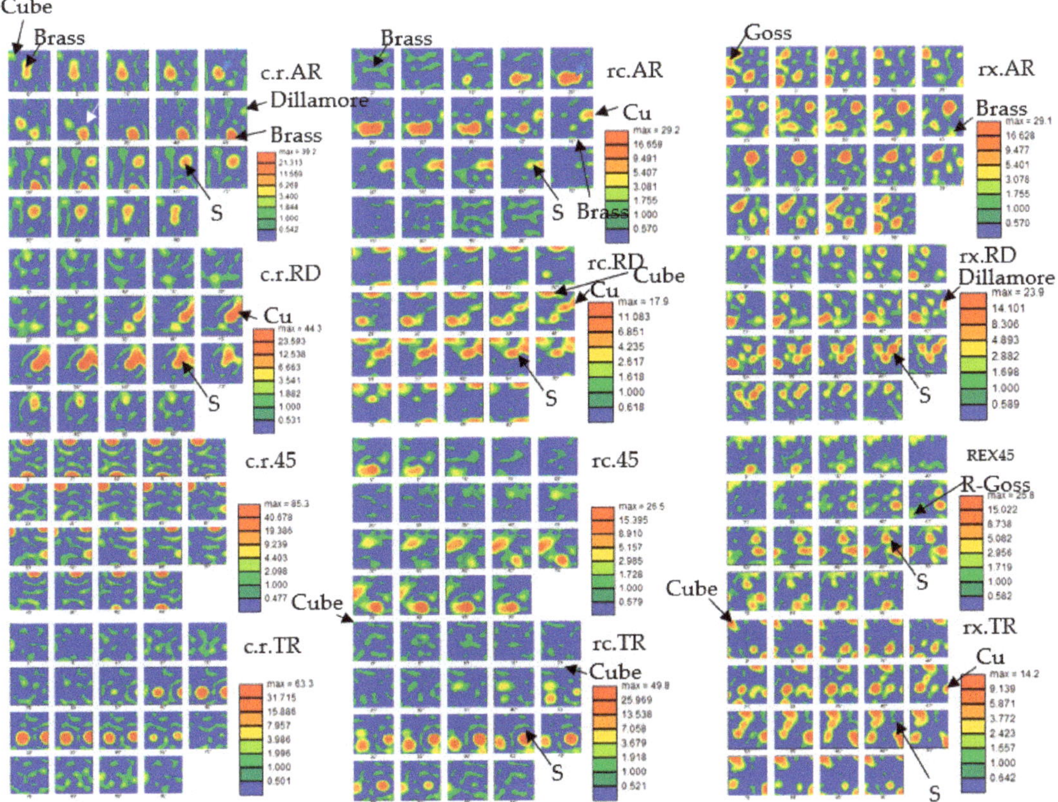

Figure 13. ODFs of the examined samples.

4. Discussion

In the present study the microstructure and texture properties of Al3104 sheet samples were evaluated by means of optical and scanning electron microscopy with EBSD in order study the correlation of the observed texture components resulting from the deformation and annealing processes with the measured r-values.

With regards to samples subjected to tensile testing, an improved behavior was reported for the recovered sample since it maintained its toughness at a significant degree and at the same time exhibited a significantly higher elongation value. This is further supported by the measured Δr and r_m values indicating an optimum response towards drawing among the samples examined. Hardness decreased due to the re-arrangement of dislocations induced during cold rolling. Electrical conductivity was elevated as the Al matrix increased through the annealing process. This was mainly due to the increased average free movement distance of the electrons and secondly due to the re-arrangement of the alloying elements within the matrix during the annealing process.

The EBSD measurements underlined the significant effect of the deformation process on different areas of the microstructure. A shift of the crystallographic orientation during thermomechanical processing is observed through the evolution of SGBs. This is the key parameter for the correlation between anisotropy, mechanical behavior, and grain orientation. The slip pattern was found to be the connecting factor of a microstructural characteristic such as the misorientation angle boundaries and the crystallographic orientation obtained during deformation [21]. The recovered state of the Al sheet was examined since dislocations and misorientation grain boundaries response in relation to tensile testing and subsequent fracture is obscure [22]. Different amounts of misorientation boundaries, differences in morphology, dislocation density, and stored energy all had a great impact on the material's response towards recovery and recrystallization. Most of the dislocations concentrated at the boundaries are induced through active slip systems. The total shear can be separated towards different slip systems and according to Taylor [23], five slip systems represent the required number of slip systems for the shape changes in an undergoing deformation. However, inside the cell block, maybe less than five slip systems could be sufficient to induce dislocation jog formation [24]. Dislocation boundaries form a subdivision of the deformation microstructure, where various characteristics are provoked through different orientations of the initial deformed grain. This correlation is important since the microstructural characteristics had a great impact on the mechanical behavior of thermomechanically treated metals [22]. The dislocation boundaries which exhibit a correlation with the slip planes {111} and a 5° difference among the slip lines and the dislocation boundaries are characteristic of deformation from a crystallographic perspective. At the same time, structure morphology, misorientation angles and dislocation boundaries indicate the presence of variations with regards to dislocation density and stored energy [22].

For Al sheets produced through cold rolling, such as those examined in the present study, the Schmid factor analysis indicates that the augmented number of available slip planes leads to changes of the orientation of the boundaries. The orientations obtained non-crystallographic positions [25], from a state of being almost parallel to slip planes. Therefore, it could be reasonable for one to correlate the microstructural characteristics and the behavior of the sheet after plastic deformation. In all examined cases, the number of active slip systems is reduced during plastic deformation to ease the deformation locally. This behavior is expected when the deformation is inhomogeneous [26], a condition not relating to slip zones. After the uniaxial loading, which results to fracture, transformations of SGBs to HAGBs were evident, in the rx.45 sample, which exhibited a homogeneous distribution of the slip zones over the crystal volumes thus it shed light to the "fragmentation" as well as the "mosaic" of SGBs into the crystal [26]. The differences with regards to the evolution of the SGBs could be attributed to neighboring regions which are suppressed in different lattice rotations.

The main structural parameters considered are the following: (i) boundary morphology, (ii) crystallographic orientation, (iii) crystallographic texture, (iv) misorientatation angle boundaries, which express the deformation of the material in different terms in comparison to the macroscopic plastic measurements in small to medium strains. In higher strains ($\varepsilon > 1$) lamellar bands (LBs) are formed. When an Al sheet is subjected to deformation, a part of the mechanical energy is stored (as dislocations mostly). Dislocations are not allocated inside the material randomly. They are concentrated at dislocation boundaries functioning as dividers of low dislocation density areas. This subdivision is common for the fcc alloys with medium to high SFE. Increased strains led to an average misorientation angle of $\geq 15°$. Parts of the grain rotate differently towards various rotations inside the grain, and consist of the observed texture components [21]. This phenomenon is observed in highly deformed grains, and is visually represented in IPF maps as a chromatic difference in in the interior of the grain. Such differences can be the result of localized shear [21,27]. From a microstructural point of view the pattern that is created due to rolling relates to the arrangement of dislocation boundaries planar cell blocks measured as geometrically necessary dislocations (GNDs)] [21]. The dislocations located at the boundaries are known as GNDs (GND Nye tensor). The formation of those dislocations relates to the lattice rotation required for the individual grains of the polycrystalline material to be plastically deformed, and convert into SGBs, [28]. Knowledge of the distribution of the misorientation boundaries could clarify the subsequent behavior observed during recrystallization, since the state of the deformed microstructure determines the process of nucleation and subsequent growth. The misorientation boundaries lead to an estimation of the stored energy [29]. Dislocations "jump over" through small SBGs easier than larger ones, leading to the assumption that smaller subgrains are harder [30,31]. This could also be understood by the fact that boundary conditions of higher order are expected from the lattice rotation of smaller SGBs. This could be related to the higher R_m accompanied by a higher amount of SGBs in sample rx.RD.

With regards to texture components, the Dillamore component relates to the Cu through a rotation of $8°$ relative to the transverse direction (TD). In the recovered state of the rc.AR sample, well-defined SGBs and thus dislocation-free grains were observed, indicating the high mobility of the dislocations as well as the fact that they are rearranged through cross slip and climb during deformation. Due to the existence of shear bands, a component Q {013} <231> forms, which appears in significantly lower texture intensities compared to cube texture. This explains the high amount of Q component observed in the cold-rolled sample. Concerning the particle stimulated nucleation (PSN) mechanism, it is important to note that the cube orientation tends to increase at the expense of the PSN orientations, such as P, and the precipitation of Mg_2Si particles tends to impede PSN more strongly than the nucleation at the cube bands [7].

Additionally, another recrystallization texture component appears in rolled and annealed aluminum sheets, namely the R {124} <211> component. It mainly appears due to the presence of Fe in the alloy and indicates either in-situ recrystallization or a strain-induced boundary migration nucleation process with subsequent growth. The R texture is usually interpreted as, a retained rolling texture S component although it exhibits distinct differences in its intensity and its exact positioning therefore their observed coexistence in the examined samples could be explained. With regards to the component's growth, it appears that the R orientation emerges and grows substantially [7]. The R component is retained from rolling texture components in the cases with reduced stored dislocation energy (at recovery state), as it observed in the rc.AR sample. The high percentage of S components was also important as these grains, favor the nucleation of R components as well [32] between the deformed bands. The balance between deformation textures and annealing textures is crucial to controlling earing in beverage can stock [10,33]. The solute Mn plays an important role in decreasing the EC value of Al alloy [34]. Therefore, the lower hardness but higher EC value can possibly be correlated with a lower content of Mn in solid solution. It has been found that the higher the intensity of the deformation Cu component, the stronger the intensity of the subsequently produced cube component. Similarly, the intensity of the Goss component is directly related to the strength of the brass component which exhibited an intense presence in the cold rolled and recovered samples. Generally, brass and copper exhibit a greater tendency to recrystallize discontinuously, whereas at these cases S is retained from the previous cold rolling states. P and Q components originate in deformation inhomogeneities, i.e., particles or shear bands [13,35]. The SGBs cannot "escape" from the particles and therefore the evolution of the substructure is controlled by subgrain growth which is directly correlated with the presence of second phase particles. The formation of a high percentage of HAGBs was observed at the rx.AR sample with a microstructure resembling a material subjected to discontinuous recrystallization. The driving force for the subgrain growth is directly proportional to the shape factor and misorientation dependent stored energy of the subgrain array and inversely proportional to the radius of the subgrains. Subgrains can grow through two mechanisms, i.e., boundary migration and coalescence of subgrains. In the case of boundary migration, subgrain coarsening occurs due to the migration of the low-angle boundaries. Large subgrains will grow at the expense of small subgrains and the exact mechanism will rely on the dislocation structure at the boundaries as well as the mobility of the triple junctions. Subgrains also grow through rotation assisted coalescence of subgrains. The recrystallized grains of the rx.AR sample exhibited quality patterns of higher quality and confidence index values from recovered grains. In addition, no orientation variation inside the grains was observed. Through the shift of the texture maximum observed at the ODF maps in the several φ_2 sections, R oriented grains exhibited a competitive behavior towards brass, copper, and S. The SBGs formed near S orientations are more favorable to grow [32].

An optimal combination of a minimum Δr and a maximum rm value was observed for the recovered state sample (rc.AR). The high amount of S components (S2 and S3), the strain-free elongated grains accompanied by an average amount of SGBs as well as the high amount of HAGBs in the as-received state, obstruct dislocations from overpassing them. The contribution of precipitates and dispersoid particles could be neglected if the precipitate spacing is greater than the mean free path of the electron, which is possibly the case as indicated by conductivity measurements. In the intermediate range of precipitate and dispersoid particle spacings commonly observed in commercial aluminum alloys, the contribution of precipitates to electrical resistivity is less clear [36–38]. A deformed material exhibiting high dislocation density initially forms cells and then exhibits annihilation of dislocations within the cells resulting into subgrain formation [30]. The increase of EC was evident in the rx.AR sample and could be attributed to the precipitation of dispersoids. The percentage of recrystallization components after each thermal treatment, increased slightly for the lower annealing temperature (rc.AR) and more dynamically for the higher ones (rx.AR). IPF intensities did not exhibit high variations (2.0 to 2.5) indicating a relatively weak overall texture. In the as-received condition, the most prominent component was S whereas after annealing, the most prominent components were R, Goss, and Q. Conventional or discontinuous dynamic recrystallization involves the nucleation of new, dislocation-free grains followed by their subsequent growth through the migration of HAGBs. Such nucleation occurs in areas where the strain and orientation gradients are high. As such, the process can be differentiated from recovery (where HAGBs do not migrate) and grain growth (where the driving force is the reduction in the boundary area). During plastic deformation, the work performed is the integral of the stress and strain in the plastic deformation regime. Although most of this work is converted to heat, some fraction (~1–5%) is retained in the material as defects—particularly dislocations.

5. Conclusions

The most significant conclusions of the presented work are summarized below:

- The optimal combination of Δr and r_m values was observed for the recovered sample.
- The S component exhibited a noteworthy behavior since, S2 and S3, namely similar components, retained the 50% of the texture before and after tensile testing in an orientation parallel to the RD.
- The texture components exhibited gradual orientation transitions over various layers of SGBs.
- The most prominent recrystallization texture component in the cold rolled and recovered states was R due to the previous intense presence of S components.

- The recovered sample exhibited a recrystallization texture leading to better results with regards to the deep drawing behavior of the material, an assumption that was further supported through the measured r-values.
- After uniaxial loading, the subgrain structure exhibited a gradual transformation to HAGBs in the recrystallized sample at 45° while a reinforcement of the SGBs was observed in all other examined cases.
- The (101) rotation towards (112) led to satisfying results regarding the resulting mechanical response.
- The combination of a high number of S-oriented grains, which were at the same time, strain-free due to recovery, as along with the presence of HAGBs and secondarily by SGBs, constituted the ideal substructure regarding the Δr and $r_{m\ values}$. The aforementioned state is expected to lead into better results with regards to the deep drawing behavior of the material.

Author Contributions: Conceptualization, S.P. (Sofia Papadopoulou); Methodology, S.P. (Sofia Papadopoulou), S.P. (Spyros Papaefthymioub); Software, S.P. (Sofia Papadopoulou); Validation, S.P. (Sofia Papadopoulou); Formal Analysis, S.P. (Sofia Papadopoulou), Investigation, S.P. (Spyros Papaefthymioub), Z.Z.; Resources, S.P. (Sofia Papadopoulou), Z.Z.; Data Curation, S.P. (Spyros Papaefthymioub); Writing—Original Draft Preparation, V.L.; Writing—Review and Editing, S.P. (Sofia Papadopoulou) and S.P. (Spyros Papaefthymioub); Visualization, S.P. (Spyros Papaefthymioub); Supervision, S.P. (Sofia Papadopoulou) and S.P. (Spyros Papaefthymioub); Project Administration, S.P. (Sofia Papadopoulou). All authors have read and agreed to the published version of the manuscript.

Funding: This research received no external funding.

Institutional Review Board Statement: Not applicable.

Informed Consent Statement: Not applicable.

Data Availability Statement: Raw data are restored in the relevant electronic folders of the testing equipment.

Acknowledgments: The authors would like to express their gratitude to ELKEME S.A. and ELVAL S.A. managements for all the kind support.

Conflicts of Interest: The authors declare no conflict of interest.

References

1. Charlier, H.; Ekström, P. *Aluminum Alloys for Packaging II*; Institute of Materials Engineering Australasia Ltd.: Victoria, Australia, 1996; pp. 245–251.
2. Kaufman, J.G.; Anderson, K.; Weritz, J. *Properties and Selection of Aluminum Alloys*; ASM International: Materials Park, OH, USA, 2019.
3. *ASTM E407-07 Standard Practice for Microetching Metals and Alloys*; ASTM International: Materials Park, OH, USA, 2015.
4. Benke, M.; Hlavacs, A.; Petho, D.; Angel, D.; Sepsi, M.; Mertinger, V. A simple correlation between texture and earing. *IOP Conf. Ser. Mater. Sci. Eng.* **2018**, *426*, 012003. [CrossRef]
5. Aretz, H.; Aegerte, J.; Engler, O. Analysis of earing in deep drawn cups. *AIP Conf.* **2010**, *1252*, 417.
6. Tikhovskiy, D.F.I. Simulation of earing during deep drawing of an Al–3% Mg alloy (AA 5754) using a texture component crystal plasticity FEM. *J. Mater. Process. Technol.* **2007**, *183*, 169–175. [CrossRef]
7. Zhao, Z.; Mao, W.; Roters, F.; Raabe, D. A texture optimization study for minimum earing in aluminium by use of a texture component crystal plasticity finite element method. *Acta Mater.* **2004**, *52*, 1003–1012. [CrossRef]
8. *OIM Analysis 5.3 Manual*; OIM: Le Grand-Saconnex, Switzerland.
9. Barros, P.; Oliveira, M.; Alves, J.; Menezes, L. Earing prediction in drawing and ironing processes using an advanced yield criterion. *Key Eng. Mater.* **2013**, *554–557*, 2266–2276. [CrossRef]
10. Kocks, U.F.; Tome, C.N.; Wenk, H.R. *Texture and Anisotropy*; Cambridge University Press: Cambridge, UK, 2000.
11. Wert, J.A.; Huang, X.; Winther, G.; Pantleon, W.; Poulsen, H.F. Revealing deformation microstructures. *Mater. Today* **2007**, *10*, 24–32. [CrossRef]
12. Liu, W.; Jensen, D.J.; Morris, J. Effect of grain orientation on microstructures during hot deformation of AA 3104 aluminium alloy by plane strain compression. *Acta Mater.* **2001**, *49*, 3347–3367. [CrossRef]
13. Hirsch, J.; Engler, O. Texture, local orientation and microstructure in industrial Al-alloys. In Proceedings of the 16th RISO International Symposium on Materials Science, Roskilde, Danmark, 4–8 September 1995.

14. Hirsch, J.; Al-Samman, T. Superior light metals by texture engineering: Optimized aluminum and magnesium alloys for automotive applications. *Acta Mater.* **2013**, *61*, 818–843. [CrossRef]
15. Huang, K.; Li, Y.; Marthinsen, K. Effect of heterogeneously distributed pre-existing dispersoids on the recrystallization behavior of a cold-rolled Al-Mn-Fe-Si alloy. *Mater. Charact.* **2015**, *102*, 92–97. [CrossRef]
16. Daaland, O.; Nes, E. Origin of Cube texture during hot rolling of commercial Al-Mn-Mg alloys. *Acta Mater.* **1995**, *44*, 1389–1411. [CrossRef]
17. Worswick, M.; Finn, M. The numerical simulation of stretch flange forming. *Int. J. Plast.* **2000**, *16*, 701–720. [CrossRef]
18. Voort, G.F.V. *Metallography, Principles and Practice*; McGraw-Hill: New York, NY, USA, 1984.
19. Takeda, H.; Hibino, A.; Takata, K. Influence of Crystal Orientations on the Bendability of an Al-Mg-Si Alloy. *Mater. Trans.* **2010**, *51*, 614–619. [CrossRef]
20. Lehto, P. Adaptive domain misorientation approach for the EBSD measurement of deformation induced dislocation sub-structures. *Ultramicroscopy* **2021**, *222*, 113203. [CrossRef] [PubMed]
21. Hansen, N. New discoveries in deformed metals. *Met. Mater. Trans. A* **2001**, *32*, 2917–2935. [CrossRef]
22. Liu, Q.; Jensen, D.J.; Hansen, N. Effect of grain orientation on deformation structure in cold-rolled polycrystalline aluminium. *Acta Mater.* **1998**, *46*, 5819–5838. [CrossRef]
23. Taylor, G. Plastic Strain in Metals. *J. Inst. Met.* **1938**, *62*, 307–324.
24. Bay, B.; Hansen, N.; Hughes, D.; Kuhlmann-Wilsdorf, D. Evolution of fcc deformation structures in polyslip. *Acta Metall. Et Mater.* **1992**, *40*, 205–219. [CrossRef]
25. Liu, Q.; Hansen, N. Deformation microstructure and orientation of FCC crystals. *Phys. Status Solidi* **1995**, *149*, 187–199. [CrossRef]
26. Sedlacek, R.; Blum, W.; Kratochvil, J.; Forest, S. Subgrain Formation during Deformation: Physical Origin and Consequences. *Metall. Mater. Trans. A* **2001**, *32*, 319–327.
27. Hansen, N.; Jensen, D.J. Development of Microstructure in FCC Metals during Cold Work. *Philos. Trans. Math. Phys. Eng. Sci.* **1999**, *357*, 1447–1469. [CrossRef]
28. Ashby, M. The Effect of Dispersed Phases upon Dislocation Distributions in Plastically Deformed. *Philos. Mag.* **1996**, *13*, 805–834.
29. Godfrey, A.; Mishin, O.; Yu, T. Characterization and influence of deformation. *IOP Conf. Ser. Mater. Sci. Eng.* **2015**, *29*, 012003. [CrossRef]
30. Humphreys, F.; Hatherly, M. *Recrystallization and Related Annealing Phenomena*; Elsevier: Amsterdam, The Netherlands, 2004.
31. Humphreys, F. The nucleation of recrystallization at second phase particles in deformed aluminum. *Acta Metall.* **1977**, *25*, 1323–1344. [CrossRef]
32. Engler, O. On the origin of the R orientations in the recrystallization textures of aluminum alloys. *Met. Mater. Trans. A* **1999**, *30*, 1517–1527. [CrossRef]
33. Papadopoulou, S.; Kontopoulou, A.; Gavalas, E.; Papaefthymiou, S. The Effects of Reduction and Thermal Treatment on the Recrystallization and Crystallographic Texture Evolution of 5182 Aluminum Alloy. *Metals* **2020**, *10*, 1380. [CrossRef]
34. Li, Y.; Arnberg, L. Quantitative study on the precipitation behavior of dispersoids in DC-cast AA3003 alloy during heating and homogenization. *Acta Mater.* **2003**, *51*, 3415–3428. [CrossRef]
35. Papadopoulou, S.; Gavalas, E.; Papaefthymiou, S. Methodology for the Identification of Nucleation Sites in Aluminum Alloy by Use of Misorientatation Mapping. *Mater. Proc.* **2021**, *3*, 11. [CrossRef]
36. Rossiter, P. *The Electrical Resistivity of Metals and Alloys*; Cambridge University Press: Cambridge, UK, 1987.
37. Muggerud, A.; Mørtsell, E.; Li, Y.; Holmestad, R. Dispersoid strengthening in AA3xxx alloys with varying Mn and Si content during annealing at low temperatures. *Mater. Sci. Eng. A* **2013**, *567*, 21–28. [CrossRef]
38. Lodgaard, L.; Ryum, N. Precipitation of dispersoids containing Mn and/or Cr in Al–Si alloys. *Mater. Sci. Eng. A* **2000**, *283*, 144–152. [CrossRef]

MDPI

St. Alban-Anlage 66

4052 Basel

Switzerland

Tel. +41 61 683 77 34

Fax +41 61 302 89 18

www.mdpi.com

Metals Editorial Office

E-mail: metals@mdpi.com

www.mdpi.com/journal/metals